Right-Brained Place Value

Adding & Subtracting Numbers Over 10

Multiple-digit addition and subtraction made easy

---

a Forget Memorization book

Effortless learning through images, stories, hands-on activities, and patterns

by Sarah Major

www.child1st.com

To request more information regarding the copyright policy, contact:

Child1st Publications
PO Box 150226
Grand Rapids, MI 49515
800-881-0912

info@child1st.com

For other teaching and learning resources designed for visual, tactile, kinesthetic and other right-brained learners, visit www.child1st.com.

Other books and materials by this author:

I Can Sing from 1 to 10
Right-Brained Addition & Subtraction
Right-Brained Time, Money & Measurement
Right-Brained Multiplication & Division
Right-Brained Fractions
The Illustrated Book of Sounds & Their Spelling Patterns
The Easy-for-Me™ Children's Readers
SnapWords® Sight Words Cards
Alphabet Tales
Stylized Alphabet
SnapWords® Spelling Dictionary, 2nd Edition

ABOUT THIS BOOK

This book is for children who are strongly visual, who learn all at once through pictures, are drawn to patterns, rely on body motions, and who need to understand the process behind each math problem they solve. Child1st teaching and learning resources all follow the principle of conveying teaching using a variety of right-brain-friendly elements. We take learning concepts that utilize symbols (numbers and letters) and abstractions, which are left-brained, and embed them in right-brained elements to beautifully integrate the left and right hemispheres in the brain.

Right-brained Elements:

1- We embed symbols in VISUALS so that the child can take a quick look, absorb the learning piece, and store it as an image to be retrieved intact later.

2- We use PERSONIFICATION which is a powerful element in teaching and learning. The use of personification makes for rapid learning because the very look and personality of the character conveys the substance of the learning. For example, in this book, dot patterns take on the characters of a bear's face, a high chair, a row of bushes, etc.

3- We rely on PATTERN DISCOVERY as a way of making numbers come alive and as a means of conveying the amazing relationships between numbers. What results is number sense. Because the brain is a pattern seeking organ, it is drawn to material that follows patterns. It is my desire that through this teaching resource, many children who are overwhelmed or daunted by math might come to truly be fascinated by it instead.

4- We use STORY to contain and convey the meaning of what we are teaching. Stories, like visuals, make learning unforgettable. They explain the "why" behind math concepts and tie everything together, creating a vehicle for meaning and for recall.

5- We use BODY MOTION—both gesture and whole body movement that mirrors the symbol shape or the action in the math story (such as addition or subtraction). Again, body movement is a powerful agent for learning and remembering. For many people, body motion makes recall effortless if the idea in the lesson is directly tied to a unique motion.

6- We employ VISUALIZATION—a powerful tool for right-brain-dominant learners. If these learners are given time to transfer the image on the paper in front of them to their brains (prompt them to close their eyes and SEE it in their mind's eye), they will be able to retrieve that image later. If the image contains learning concepts, this is how they will remember what you want them to learn. So in this book, each time a visual is introduced, prompt the student(s) to "see" the image in their mind, eyes closed.

HOW TO USE THIS BOOK

Because this book builds on *Right-Brained Addition & Subtraction* (or Book 1), please familiarize yourself with Chapters 2-4 of that book first, including "Good Practice," "Assessments," "Visual Imprinting," and "Learning Numbers." You should also be familiar (from Book 1) with the 5-Frames, "My Two Hands," and Stony Brook Village.

For students just finishing Book 1, who are fluent with computation to 10, skip to Chapter 3 to begin new material.

For students unfamiliar with Right-Brained computation, teach Book 1 Chapters 5-7 before beginning this book.

For students needing a review of Book 1, skip to Chapter 2 for a quick review of Book 1. Make sure the children are completely fluent with computation to 10 before beginning this book.

For remediation, determine whether the students lack skills in computation to 10. If they do, use Chapter 2 of this book first, or if they need more practice, use Book 1, Chapters 5-7.

GLOBAL VIEW of Right-Brained Place Value

The following table outlines the contents of this book, with a brief explanation of the focus of each chapter.

Chapter	Content
Chapter 1: *How Children Learn*	**Who is targeted and why:** Identifying the children who have difficulty with math and ways to address this difficulty.
Chapter 2: *Prerequisite Skills*	**Prior knowledge:** Identifying students' background knowledge and reviewing foundational concepts before moving into multi-digit computation.
Chapter 3: *Place Value*	**Teaching place value to young ones:** Presenting a hands-on, whole-body method of teaching place value that draws on real-world applications.
Chapter 4: *Computation Using Place-Value Mats*	**Practicing place value on mats:** Using real materials as a class or in small groups to gain fluency in understanding and using place value.
Chapter 5: *A Bird's-Eye View*	**Looking globally at all problems through 20:** Identifying problems and their place within the global whole of computation over 10; working with problems children already know; identifying problems yet to be mastered.
Chapter 6: *Make a 10*	**Learning a method of addition over 10 based on place value:** Working with addition problems that require making a ten, using real materials and building on prior knowledge of computation to 10.
Chapter 7: *Take from 10*	**Learning a method of subtraction based on place value:** Working with subtraction problems that require taking a ten, using real materials and building on prior knowledge of computation to 10.
Chapter 8: *Taking Stock*	**Assessing fluency with new concepts:** Thoroughly reviewing all newly learned material to determine mastery.
Chapter 9: *To the Top!*	**Extending the method for all multi-digit numbers:** Using this method for all numbers, for those students who want to learn more.
Appendix A:	**Activities & Resources:** All materials necessary to implement the lessons in the book.
Appendix B:	**Monitoring forms:** Reproducible forms for tracking individual students and whole-class progress, geared to the content of each chapter.
Appendix C:	**Answer Keys**

TABLE OF CONTENTS

CHART of LEARNING STYLES

Traditional Methods are	
designed for these learners	**but not for these learners!**

How do I learn? (Dr. Anthony F. Gregorc)

I perceive the world

Concrete
I use my senses to take in data about the world.
What I see is what is real to me.

Abstract
I visualize, intuit, imagine, read between the lines, and make connections.
I pick up subtle clues.

I order the information I perceive

Sequential
I organize my thoughts in a linear, step-by-step manner.
I prefer to follow a plan.

Random
I organize my thoughts in segments. I will probably skip details and even whole steps, but I will still reach the goal. I like to make up my own steps.

How do I remember? (Raymond Swassing & Walter Barbe)

Visual
I need to see it. I make visual associations, mental maps or pictures, and see patterns.

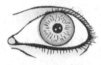

Auditory
I listen to directions.
I need to hear the sounds.

Kinesthetic
I remember well what I learn through my body. I learn best by actually doing the job.

How do I understand? (Herman Witkin)

Analytic
I am good with details, can follow steps and hear instructions, and like to finish one thing at a time.

Global
Show me the big picture! I need to see how all the parts fit in. I can hear directions after you show me the goal.

How am I smart? (Howard Gardner)

Verbal/Linguistic
I am verbal! I can speak, write, debate, and express myself well through words. IQ tests love me!

Visual/Spatial
Show me a map and I'll have it! I make vivid mental images and can use these to recall associated information. I want to see how something fits into its environment or surroundings.

Logical/Mathematical
I rely heavily on my logic and reasoning to work through problems. I am a whiz on standardized tests.

Body/Kinesthetic
I combine thinking with movement. I do well with activities that require precise motions. I learn by doing; my attention follows my movements.

Global View of Computation to 10 (from Book 1)

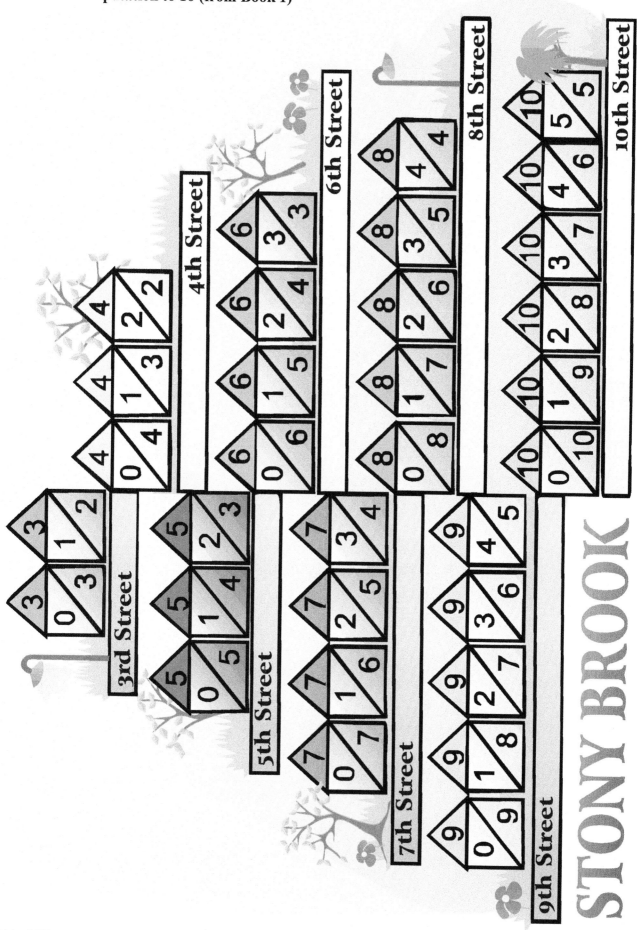

8

INTRODUCTION

My own struggles with math during my elementary years fueled my passion to discover ways of teaching that would result in success for all students.

I grew up saying and therefore believing that I just "didn't have a mathematical mind." I've since learned that my strongly visual and right-brained approach to taking in and processing information is a great enhancement to the understanding of math, *if* I utilize my visual mind in the learning and doing of math.

In graduate school, I experienced full-blown panic when I had to take my math methods class. The only way I could make sense of the problems was to draw pictures of what was going on in them. My panic skyrocketed the night my professor leaned over where I sat hunched, trying to hide the little pictures I was drawing from her view. In spite of my best attempts to cover my paper with both arms, her sharp eyes spied the drawings and she snatched my paper up. Waving it in the air, she bellowed, "Look at what Sarah's doing!" I was mortified, my body chilled, but my face burned and my stomach clenched so tightly I thought I would be sick.

What happened next surprised me. The professor went on to say that the ability to visualize math and to draw pictures of what was happening would make me a good teacher, because most children are very visual and learn better with the use of visuals. She took what to me had always been an embarrassment and pronounced it good. I will never forget that night!

During the past several years, I have worked closely with a wide variety of learners, earnestly desiring to understand what they needed in order to learn efficiently. I made it my mission to figure out the ways children learn best by throwing out any procedure that did not work for all of the students or that shut some of them down, did not hold their interest, or intimidated them in any way. Over time, I learned that there are some basic elements of practice that work with all my students regardless of their individual learning styles.

The methods in the Right-Brained Computation series are the result of working with children while focusing on the question, *"What will make it possible for **all** of them to learn?"* This quest intensified as I worked on my graduate studies in education. At the same time as I worked through content and learning challenges in my "lab" with the children, I was reading about the same learning issues in my classes. The result for me was maximized understanding. I was primarily observing the children, and then I would go to class to try and find explanations for what I saw rather than simply learning from lecture and books and then walking into the classroom to teach from that knowledge. I want to share with you a little bit of what I came away with in the hope that something in this series will support you as you give all your students the experience of success.

I am grateful to the people who together supported this work: my own children, who cheered me on; my students' parents, who believed in me and in their own children; and the professors who let me concentrate on the topics that really interested me. To the children I have worked with, all I can say is, "Thanks for letting me learn from you! You are the best!"

~Sarah Major

1 HOW CHILDREN LEARN

To fully understand the value of this method of teaching computation, it is helpful to take the time to look globally at learning styles and how they relate to traditional ways of teaching math. This overview will lead naturally into identifying the factors that are necessary for a teaching approach to be successful for all children, regardless of their learning needs.

The idea of teaching to multiple learning styles could be intimidating. Because this is true, I first provide an orientation to the learning needs children have, and then pull all the ideas together and draw some conclusions that will help to bring the seemingly disparate elements together into one simple plan for good teaching practice.

The chart on page 6 presents an overview of the learning styles that are critical to this discussion. (For in-depth information on these and other learning styles, please refer to Barbe 1985; Gardner 1993; Gregorc 1982; Tobias 1994; and Witkin 1977.) If we imagine that the learning styles on the right side of the chart represent real children in real classrooms, it will become easier to see which children are being "taught around" in traditional methods of teaching.

LEARNING STYLES IN TRADITIONAL METHODS OF TEACHING MATH

Math is normally taught in tiny steps; students are given seemingly unrelated bits of information to work with or are given steps to memorize for solving problems, typically without time for the student to discover why those steps work. Often there is no real-life application within the problems, as students work solely with paper and pencil, and miss the opportunity to construct meaning for themselves using real objects.

Is it possible to teach math, as seemingly concrete and sequential as it is, in a way that will reach the abstract, random, visual/spatial, kinesthetic, and global students? Or should we continue expecting them to learn in a left-brain manner? It seems more reasonable to change our current practice to fit the children rather than trying to force children to be something they cannot be. Let's make the assumption, then, that we should expand our method of teaching math to encompass and embrace all our students. What we will do in this book is approach computation in a global, visual, kinesthetic, abstract, and random way so that no child is left out!

THE COMMON DENOMINATOR

I believe that children who are highly visual also tend to be global, somewhat random, and kinesthetic. Visual children see a whole picture, see smaller elements within the whole picture, see their connection to other elements in the whole picture, and tend to remember parts of the picture based on where each part fits into the whole. In addition, highly visual children will move randomly through the picture (or map or pattern) and are often inclined to spatial activities that require tactile skill. Visual children will prefer to see the task done as they learn it, rather than hearing it explained, and will profit from doing the problem themselves. It is important for these children to understand why the formula works by seeing how one action affects other parts within the whole. They might not understand the process the way another student does, but if they are certain of the goal of the lesson, they will invent good steps that make sense to them and allow them to reach the goal.

Unique Features of the Kid-Friendly Computation Method

1. Students are able to add and subtract multi-digit numbers using the same strategies they learned in Book 1 for computing to ten, including "My Two Hands." For this reason, I studiously avoid terms such as "borrowing" or "carrying." Every step I do is based on adding numbers under ten or subtracting numbers from ten. Carefully limiting the procedures to these basic facts ensures students will be successful in computation with numbers of any size.

2. A critical feature is that rote memorization—whether of answers to problems, steps in solving a problem, or procedures for computation— is avoided at all costs. Memorization is not the answer. Drill and repetition of facts might result in short-term memorization for some students, but not for every one. If you take the time to tie a process, fact, or procedure to a story, long-term recall is much more likely because the procedure is tied to a hook that acts as a trigger for recall.

3. In this book, the actions of adding and subtracting are tied directly to stories from the beginning so

that students are able to "see" computation taking place. As a result, they not only understand and remember what is going on, but they also can easily determine which action to use when they are presented with story problems.

4. Computation is introduced using concrete materials in order to add a visual/tactile component, which will boost learning. Having children do the actions of computation using real materials will also cement the procedures in their minds.

5. The reason for the success of the methods in the Right-Brained Computation series is due to the fact that they tap auditory, kinesthetic/tactile, and visual components simultaneously. Whatever I teach, I teach to the ears, eyes, and hands/body using as many vehicles as possible to carry learning concepts into the brain.

6. Assessments are an important component of this program. I believe that the purpose of teaching is to provide all children with the opportunity to master the material and that all children are capable of mastery. Assessments are simply activities that reflect back to us as teachers what we have successfully taught and where reteaching needs to happen. My assessments are often reproductions of worksheets the children have completed during daily lessons which reflect whether the children indeed know the material I have taught them. There is no point, then, in presenting the material in a different format. The children either will or will not know how to do the computation. I often recommend giving a second assessment in order to rule out the possibility that a child who did not do well was simply tired, distracted, or having an "off" day.

7. If a child is unable to correctly complete a portion of the material in an assessment, this is a signal that reteaching needs to occur. Assessments are tools for determining mastery and areas needing reteaching, not means by which to level or sort students. The shift in our thinking, from "grading kids," to leading them to complete mastery of the material, leads to incredible gains in learning. Our belief in our students will result in their belief in themselves, and the outcome of that belief is success without a doubt!

8. I allow, even encourage, students to retake the same test after more practice if they are not satisfied with their performance. I began doing so because many of the children I teach have failed so often that they have given up trying or believing in their ability to learn. The results were unexpected, however. I've had students ask to retake a test not because they didn't pass, but because they wanted to prove it to themselves that they could ace it again, and also because they wanted to score 100%. Recently, after scoring 100% on their test in my classroom, my fourth graders decided to retake the test in their regular classrooms "just for fun." I'm not kidding; those were their words. Allowing retakes not only increases mastery, it (a) increases confidence, (b) raises the students' expectations of themselves from simply passing to becoming fluent and confident with the material, and (c) encourages students to take ownership of their own learning.

PUTTING IT ALL TOGETHER

Now let's take these ideas and distill from them some basic elements of a good teaching approach that will engage children from both sides of the learning styles chart:

1. **State the goal first**. Explain what the children are going to learn and why it is important. What is the point, the bottom line?

2. **Provide concrete materials** for the children and establish clear but general parameters within

which they will work. Then let them manipulate the materials until they can see the action that occurs in a computation and how that action affects the whole.

3. **Discuss with the students what they have discovered** and guide them in drawing conclusions. This step involves pattern detection and exploration. It will be through patterns that they recall specific facts.

4. **Present real-life examples** of using these sums. Use stories whenever you possibly can. Stories help demonstrate to students how math is relevant. They help answer the inevitable question, "When am I ever going to need to know this?"

5. **Allow as much practice** in solving problems as the students need.

6. Don't expect the students to "just remember" anything. Instead, **tie every new concept to a previously learned concept**, using visual and movement cues.

7. **Teach to all three modalities.** But no worries; this book will guide you in that endeavor!

PREREQUISITE SKILLS

GOALS FOR THIS CHAPTER

1. To review the meaning of numbers
2. To review the relationships between numbers
3. To review "My Two Hands," 5-frames, and dot cards
4. To review the number houses to ten

It is important to begin the lessons in this book only after making absolutely sure that all your students are fluent with the number facts of computation to 10. For example, if you show a child the numbers 2 and 5, they should be able to instantly supply the third number in that trio - 7! If 4 and 5, they will supply 9, etc.

The main headings in this chapter correspond to the areas of prior knowledge needed for multi-digit computation. Under each heading are suggested activities. If students have mastered this content, the lessons in this book will go very smoothly for them. To learn multi-digit computation, you will be taking those earlier skills, refining them, and applying them to new situations. For this reason, it is critical to ensure that they are fluent with these background concepts. Even if your students have studied Book 1 (*Right-Brained Addition & Subtraction*) in a prior year, a review will refresh their memories.

THE MEANING OF NUMBERS

Begin by laying a visual background for the numbers to ten. The value of visual imprinting is discussed in depth in Book 1, Chapter 4. Your goal is for the children to have a visual image of the quantity each number represents. Using dots to represent the numbers is an effective way to accomplish this (see below).

Discovery

Locate the dot cards (2.1, page 68) and photocopy them onto transparency film for use with an overhead projector. Also photocopy several sets on paper or card stock and laminate them for use in math centers.

Display the progression of dots from one to ten:

1	2	3	4	5	6	7	8	9	10

Ask the children to talk about what they notice about the numbers. Which groups of dots have similar shapes? Which numbers have an "odd guy" on top? Discover and compare how many pairs of dots each column number has. Identify which numbers have the same number of dot pairs (2 and 3, 4 and 5, and so on). Ask: How many odd numbers are there in our chart? (5) How many even? (5) Why do you suppose that is? How many numbers have a four-dot pattern in them? (7) How many numbers include a three dot patterns? (all odd numbers)

Number-Recognition Activities

Use the laminated dot cards for games, which children can play independently in the math center.

War: Students play in pairs. Players display a card simultaneously, facing up on the table. In the traditional game, the player who displays the higher number gets both cards. The child has to correctly name both quantities laid down in order to take the cards (without counting dots!). For variations, have the lower quantity win, or odd over even, and so on.

Memory: A small group can play memory using two sets of cards. Again, the children should name the quantities they turn over without counting the dots.

Go fish: Using four sets of cards for a small group, have children play Go Fish. Each player has to ask for the desired card by number name, not by showing the card to the group.

Name that number: Students work together in pairs, using the dot cards. One child holds up a card and the other names it as quickly as possible. Have students trade off roles of holding up and naming cards.

RELATIONSHIPS BETWEEN NUMBERS

For number relationships, begin with the 5-frame number chart (see 2.2 on page 69). You can photocopy enough copies for each child to have one, or project one copy on an overhead projector. You may wish to laminate for durability.

Begin by displaying that filled-in 5-frame chart for group discovery time (See below). Ask the children to talk about the patterns they see. Where and how do numbers repeat? Find numeral patterns such as 1, 6, 1, 6, 1 and 2, 7, 2, 7, 2 and 3, 8, 3, 8, 3, and so forth. Examine the fives column and notice how the numbers in the 10s position repeat (**10, 15, 20, 25, 30, 35, 40, 45**, etc.). What is significant about this pattern?

Number-Recognition Activities

Full frame: Give each child a blank 5-frame chart to 20 (2.3, page 70). Call out numbers at random, asking students to write each number in the box where it belongs. This activity is self-assessing, because when you are finished, the children can check their own work to see if their numbers appear in the correct sequence.

5-frame patterns: Again using the blank 5-frame chart (2.3, page 70), call out questions that require children to recognize the patterns within the chart. For example, "Write the number under eight" or "Write the number above six." This activity will help cement in students' minds the relationships between the numbers in each column of the chart.

Children who are highly visual will recognize additional patterns as they continue to use this chart. For example, they might note that adding five to any number yields the number directly underneath it on the chart. If they add ten, they will skip down two rows. If they add four, they will skip down one row then move one space to the left. To add six, they move down one row then move one space to the right.

1	2	3	4	5
6	7	8	9	10
11	12	13	14	15
16	17	18	19	20
21	22	23	24	25
26	27	28	29	30
31	32	33	34	35
36	37	38	39	40
41	42	43	44	45
46	47	48	49	50
51	52	53	54	55
56	57	58	59	60
61	62	63	64	65
66	67	68	69	70

How high can you go?: Make numerous copies of the long blank 5-frame chart (2.4, page 71) and cut apart along the dotted line. Give each child several charts to tape together end-to-end to form a really long chart. (Each child can choose how long to make his or her chart.) Then, the children write in all the numbers to fill their charts. The self-assessment piece consists of checking that patterns are correct.

5-Frame Dot Cards:

Use the 5-frame dot cards (activities 2.5a-b, pages 72-73) to help children visualize where each number falls in relationship to the base of ten. The number at the end of each full row (here, 5, 10, 15, 20) becomes an "anchor number" to which numbers are added or subtracted to yield quantities in partial rows. This visual imprinting allows children to see computation globally, a particular benefit to those who are not logical-sequential thinkers. You can also use the cards to practice oral computation. Have the student begin by naming the quantity shown (e.g., 12). Then ask questions such as, "How many more dots would you need to make 15?" "How many would you have if I gave you only two more?" "What about if I took away four?" Students can also play the card games listed on pages 16-17, such as Go Fish or Memory.

VISUAL IMPRINTING FOR COMPUTATION

Visual imprinting refers to the practice of purposeful transfer of an image to the brain by asking the child to close his eyes and see the image in his imagination. Give each student a copy of activity 2.1, the dot cards, and some colored pencils. Using the dot card transparencies you made for the "Meaning of Numbers" activities, place one card at a time on the overhead. Guide students to first of all "see" the dot pattern in their imaginations, and then take it further by recognizing subgroups within each dot pattern. Let's use these dot cards as an example to illustrate the process:

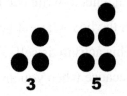

For the three-dot pattern: Ask students to first of all "see" the dot pattern in their minds. Then, find 2 + 1 in this dot pattern. Have them circle the two dots and one dot separately using their colored pencils.

For the five-dot pattern: Ask whether students can find a 4 + 1 pattern in this dot grouping. Have them circle the four dots and one dot separately. Now have them look for a 2 + 3 grouping. Using a different color, have them circle the two dots and three dots separately.

Continue in this manner with each set of dots until students have circled all the possible number combinations that add up to each number from two to ten. I have also provided sheets of hollow dots (see 2.6a-b, pages 74-75). If you would prefer, you can have children color in the subsets of dots with their colored pencils, rather than circling them. These pages can be stapled together to make a number book for each child.

My Two Hands

Now show each dot card again. As you identify the subgroupings this time, have the children show you the computation on their hands. The "My Two Hands" strategy is explained in detail in Chapter 6 of Book 1. In brief, it is a kinesthetic and visual method of showing computations by spreading the fingers apart or moving them together. Two is represented with the index and middle fingers, four with all four fingers, five with the fingers and thumb, six with the fingers and thumb of one hand plus the thumb of the other, and so on. To add

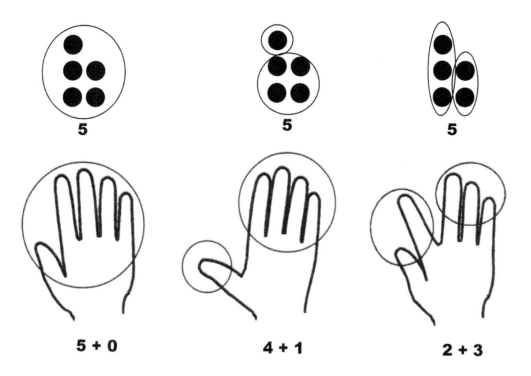

any pair of numbers totaling up to ten, students can represent each number on their fingers and simultaneously see the sum. The above illustration shows the sums for five.

Number Houses to Ten

Using activities 2.7a–2.9f (pages 76-90), follow the steps below to ensure children have mastered computations to ten. For any child who struggles with the number-house activities, back up and provide plenty of practice with the earlier activities before reintroducing number houses.

1. Refer to Chapter 7 of Book 1, if necessary, to explain the rules for filling in the houses.

2. Give each student a set of concrete objects, such as plastic counters. Have the students count out three counters, then figure out how many ways they can group them (3 + 0 and 1 + 2).

3. When students have discovered the possible groupings for 3, give them the "Figure out which families can live in each house" activity for 3s (2.7a, page 76) and have them write in the number combinations they discovered with their chips.

4. Next present the trifold activity of blank houses for Third Street (2.8a, page 80, which has two houses in each row). Have them fold under the bottom two sections, then fill in the top two houses with the correct numbers. (Each attic will have the numeral 3 in it, and the two floors will show the combinations that make up three, namely the numerals 0 and 3 in the first house and 1 and 2 in the second.) When finished, have them fold that section down and show the second row of houses. Have them complete that row in the same way, then turn to the last row of houses (thus practicing the same sums three times).

5. Repeat steps 2 through 4 for the number four (2.7b, page 76) for "Figure out which families can live in each house" and 2.8b, page 81 for practice houses).

6. Next, give students the page of mixed addition and subtraction problems for 3s and 4s to complete (see 2.9a, page 85).

7.   Repeat steps 2-6 for the numbers five and six. There is also a worksheet that mixes problems for sums from 3 to 6 (2.9c, page 87). Then repeat the process for the numbers 7 and 8. Finally, repeat the process for the numbers 9 and 10. Note that activity 2.9f has a mix of problems for sums from 7 to 10.

When the class can do all these problems fluently, move on to Chapter 3!

ASSESSMENTS

For your convenience, an individual assessment report and two whole-class record charts are provided in Appendix B beginning on page 206. They are designed for recording students' mastery of the concepts reviewed in this chapter. You can use the following assessment procedures to check for mastery of each concept. Remember that any student who has not mastered all these math skills fluently will falter as new material is introduced throughout the rest of the book. So if any students struggle, provide small-group teaching until they demonstrate readiness to move on.

Dot patterns: Mix up the set of dot card transparencies. Write down the order of the transparencies so you have an answer key to check students' papers. One at a time, show each transparency briefly. Have the children write the numeral represented by that dot pattern on their paper.

Blank 5-frame to 20: Give each student a blank 5-frame chart to 20. Call out each number from 1 to 20 in a random order (record the order for yourself) . Have students write each number you call out in its correct location in the 5-frame.

Dot flash cards: Using the 5-frame dot cards, mix them up in a random order. (Record the order for your answer key.) Flash each card briefly and have the students write down the numeral corresponding to that dot set on their papers.

Colored dot subsets: Using the pages of hollow dots, color each set of dots in two different colors to represent a pair of numbers that totals that sum (for example, five dots = 3 + 2). Hold up one of these sets of dots. Have the students represent the sum on their hands using the "My Two Hands" strategy. Do a quick visual check.

Mixed number-house problems: Give each student a sheet (or sheets) of mixed addition and subtraction problems for all the numbers to ten.

3 PLACE VALUE

GOALS FOR THIS CHAPTER

1. To act out the concept of place value
2. To apply the concept of place value to real-life situations
3. To demonstrate understanding of place value using concrete materials

Although place value traditionally is not introduced until middle elementary grades, I teach place value early as a key element of computation with multi-digit numbers. Using the visual/kinesthetic approach of Right-Brained Computation, students will first see what place value means in concrete terms, to enable them to utilize this concept as they work through the lessons in this book. Because place value is demonstrated in concrete terms using attic numerals 1 and 10, children can embrace the concept as early as kindergarten.

CONCRETE PRACTICE WITH PLACE VALUE

Just as simple sums were introduced using a story to place them in a real-world context, I extend the Stony Brook story to introduce the concept of place value.

In Stony Brook Village, just over the brook from the residential area, there is a town square with office buildings all around it. Each building is new and clean and shaded by big trees. Pointed roofs crown each building, and attic numbers (1 and 10, which represent place value) are painted on the attics like this. (Show an overhead of the empty buildings; use 3.1 on page 91.)

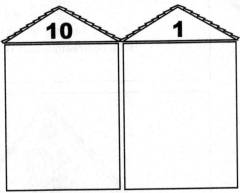

The planning commission has hired you as the property manager for one of the office buildings, and it is your job to rent office space to people who want to work inside your building. (Point to the side of the building labeled 1.) There are nine desks in this office. That means on this side you can rent desks to nine people, and no more. (Point to the side labeled 10.) On this side are big tables, each seating ten people. (Note: In this context the 10 and 1 represent place values, not a numerical total of the number of workers.)

Ten people exactly—no more, no less—must sit around each table. Once you have rented all the desks on the 1s side of the building, if more people come in looking for a place to work, you will need to take ten people from the 1s side and move them over to a big table on the 10s side. Then you can fill up the desks on the 1s side again. Each time you get ten or more people on the 1s side, you have to move a group of ten next door. If you don't, and you leave too many people on the 1s side, the commission will hear about it, and they will come with their sirens screaming to give you a ticket!

More people wanting to come in.

ACTIVITIES FOR PLACE VALUE

Act it out: Use masking tape to outline on the floor the two sides of the office, large enough that ten children can actually fit inside. (Make sure the 10s side is on the left, to mirror the order of the 10s place and 1s place in a written number.) Place nine chairs inside the 1s office and leave the 10s side empty. Tell a story in which you come to the office one day, unlock the door, and soon three children arrive, wanting to rent desks. Have three volunteers step forward, welcome them, and show them to their desks. Continue adding more occupants until the children tell you that you have reached the magic number of ten. Then have the class work out what to do next—don't tell them; let them talk it out. At this point introduce the phrase "make a ten," which you will use repeatedly throughout the method. Usher a group of ten next door, then take a large satin ribbon and gently tie the children together so they represent a group. They will not forget this object lesson! During the game, you want to reinforce the pattern of filling up the 1s side first; then, when the side fills up, making a group of ten and moving the group of ten next door. The motion is "in and to the left," which will be mirrored in written computation.

Rent-an-office game: This is a good game for children to play in the math center. Reproduce and laminate the game cards 3.2 (pages 92-93). Also photocopy the place-value mats 3.3a–b (pages 94-95). Place the two pages of 3.3 side by side and photocopy or glue onto 11" x 17" paper (making sure the 10s side is on the left). Either a pair or small group of children can play. Give the children a handful of rubber bands and a pack of wooden craft sticks. The children take turns drawing a game card and placing the corresponding number of craft sticks ("workers seeking office space") on the 1s place mat (the "office building"). As soon as the number of craft sticks exceeds nine, the children count out ten sticks, bundle them with a rubber band, and move them over to the 10s place-value mat. The purpose of this game is to practice the action of bundling ("making a ten") and then moving the bundle of ten next door. Encourage children to continue playing until they are fluent with the actions of bundling and moving sticks to represent the concept of place value.

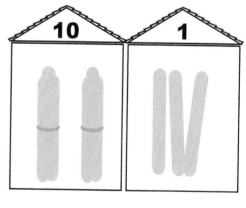

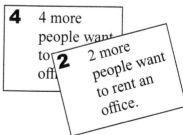

Place-Value placemats showing single and bundled sticks, and Rent-an-Office playing cards.

THE TRANSITION FROM CONCRETE TO SYMBOLIC

When you make the transition from the renting-an-office scenario to symbols, go slowly and explain what you are doing carefully, so the students do not become confused and think you are introducing a new concept. Talk out what the children have been doing in the rent-an-office game and how these actions can be represented with numbers. I like to use the place-value mat transparency and an overhead projector, but you could also draw on a whiteboard. First model for the class the action of renting out office space and making a bundle of ten using the craft sticks. Then stop and ask, "How many 10s do you have on this side?" As the children answer, write the number they say below the 10s office with a wipe-off marker. Then ask, "How many 1s do you have on this side?" and write that number below the 1s side (see illustration on the right).

Continue adding sticks to the 1s side, bundling them, and moving them over to the 10s side, until you have two bundles on the 10s side and some number of single sticks on the 1s side. Stop again and ask, "How many 10s? How many 1s?," writing the numbers below the corresponding place mats.

Next show the children a set of sticks arranged on the place-value mat and ask them to write the corresponding number. Do this several times until everyone can confidently write the correct symbols for the sets.

Last of all, show the children a number and have them draw the visual representation of that number in stick format (see illustration on the right). 3.4 (page 96) contains a set of blank place-value cards in which children can draw sticks. Repeat this step until the students have mastered it.

"What Is This?" Game

"What is this?" is a good activity for children to do in pairs in the math center. To prepare, duplicate the place-value cards 3.5a-c (pages 97-99) and laminate. These cards will have several uses in wrapping up this chapter.

1. One child holds up a card with numbers printed in the place-value houses. The second child identifies the number in terms of how many 10s and 1s it contains. For example, if the first child holds up 24, the second child would say, "Two 10s and four 1s."

2. Children can duplicate the numbers using craft sticks and the place-value mats 3.3a-b. Each child takes a turn drawing a card, then sets out sticks to represent that number on the mat.

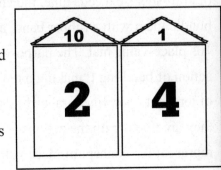

3. Children can take turns setting out sets of sticks on the place-value mats for their partners to identify using written numerals.

TAKING IT FURTHER

Inevitably, some child will ask, "What if we keep getting more and more 10s?" In this case, you should introduce the 100s office and explain that only nine bundles of ten are allowed in the 10s office. Thus, the 10s office has the same rule as the 1s office: "no more than nine." Point out that as you move left, each attic number gains one more zero. My students loved working with base ten models and filling the offices for 1s, 10s, 100s, and 1,000s. They started with the 1s, bundled a ten, added nine more 10s so they could bundle 100, and kept on going. I wrapped up by asking, "How many?" for each office and writing the number below the office. In this way, they could see that the zero means "there is no one in this office."

Because the children were interested, we continued the exploration of the four offices. I set up a model using craft sticks, and the children wrote the number it represented. Then I changed the model, and they changed their number to reflect the new model. There is something about really big numbers that fascinates children, and this activity makes large numbers understandable and accessible to them.

ASSESSMENTS

An individual progress report and a whole-class record listing the skills covered in chapters 3 and 4 can be found in Appendix B (pages 210-211). Following are suggested activities for assessment.

Give me a … : Call out a number and have the children write that number (in numerals) inside the appropriate office. (Use 3.6 on page 100; an example is preprinted on the worksheet.)

From sticks to numbers: Show a set of craft sticks in a place-value house and have the children write the number it represents. A prepared worksheet is provided (3.7 on page 101), or you could set up models using transparencies and craft sticks projected on an overhead projector.

10s and 1s: Call out a number as sets of 10s and 1s; for example, "two 10s and five 1s." Have the children write the corresponding number in numerals on a sheet of paper.

This number looks like … : Give each child a copy of 3.8 (page 102), which has blank place-value houses with numerals written underneath. The child draws sets of sticks to represent the numerals written below each house.

COMPUTATION USING PLACE-VALUE MATS

GOALS FOR THIS CHAPTER

1. To comprehend the meaning behind computation through representing it with actions
2. To become fluent in the process of computation
3. To learn to "make a ten" and "take from ten" in addition & subtraction

The purpose of this chapter is twofold. The first is for children to "see, hear, and do" the action of computing using place value, so that they comprehend the action and meaning of computation. Second, we want them to become fluent in using place value, so that when they begin solving problems, they will not be distracted by having to concentrate on how to move through the steps of computation. The best way to accomplish both of these goals is to involve the students in action and story right from the start. The stories will provide a framework that explains why one does these steps and why they work, gives relevance to the practice, and provides a procedural memory prompt.

Every child will need place-value mats for ten and one (3.3a-b), 19 sticks, and a rubber band. As you tell the following story, the children will add and subtract sticks ("people") from their mats ("offices").

STORY FOR "MAKE A TEN"

You are the property manager for your offices. You drive up on the first morning, park your car, and unlock the front door of the 1s side because the 10s side does not have a door. You walk inside and check the offices. There are nine desks on the 1s side and one conference table on the 10s side that seats ten people. More tables are stacked in the closet for later. Everything is neat and clean. You are satisfied and proud.

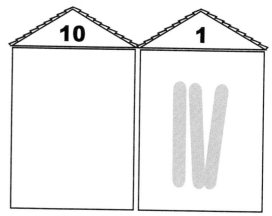

Just then, the bell jingles on the door. You hurry over and see three people standing there. They shake your hand and tell you they want to rent office space. So you show them to three desks. (Have the children pick up three sticks and place them in the 1s office.)

Just as you finish settling in your new people, the door opens and four more people walk in! (Have the children pick up four sticks.) Say, "There are three people and four more are coming. Do we have room in the 1s side for them?" You will determine that you do have room as 3 + 4 = 7.

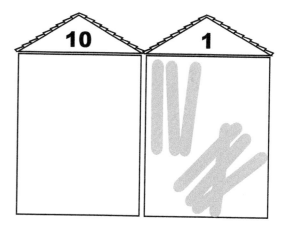

The bell jingles and four more people come in, smiling and shaking your hand. They want to rent space as well. (Have the children hold the four new sticks in their hand.) Wait a second to see if anyone says anything. If none of the children remark that will be too many people in the 1s office, say, "Will these four new people fit in the 1s office with the seven people already in there?" (No. We will need to "make a ten.")

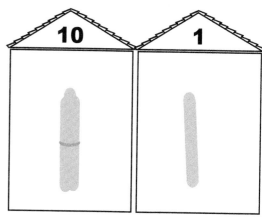

Ask, "If I have four sticks in my hand, how many should I pick up from the 1s office to make a ten?" They will need to pick up six more sticks (4 + 6 = 10), rubber band them together, and place them in the 10s office. Emphasize the phrase "make a ten." (For a visual for make a ten, see page 34.)

Again reassess the arrangement of the sticks. How many people are left on the 1s side after the children make a ten? Guide students to recognize that there is ten and one more (10 + 1 = 11). Pause here to emphasize the "in and to the left" movement, which is a movement you want to ingrain in them as they learn to add numbers with sums larger than 10. First they check the 1s side to determine if they need to make a ten or not, and if they do, they move to the left and place the ten. When they say the number, however, they will name the 10s first, then the 1s. The movement is like the shape of a backwards numeral 7. Illustrate by drawing on the board or on your overhead, as shown in the picture to the left.

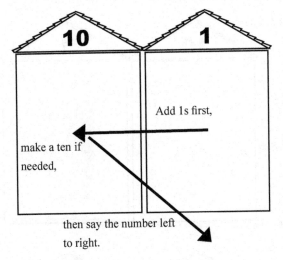

Now continue the story: "I hear the bell again! This time there are five people coming in the door! Where will you put them? Is there room on the 1s side?" Then lead them to reassess the number in the offices now to see if they need to "make a ten" (No. 11 + 5 = 16).

Continue the story, modeling simple addition through the actions of "come in, make a ten, and move to the left" until you observe that the children no longer stop to ponder where to put their "people" and that they make 10s automatically. When they can do this confidently, it is time to move to the next section.

Activities for Math Centers

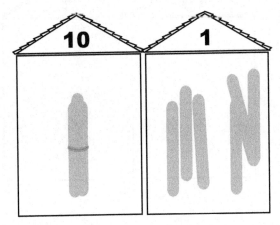

Ten and more dot cards: Copy double-sided activity 4.1 (pages 103-104). Have children quiz each other about what number is represented by each dot card. (One child holds up a card, and the other child gives the answer.) For example, for 15, the child would say, "One 10 and five 1s." (Note that the ten shaded dots represent one bundle of ten.) If you photocopy the numbers on the back side of the dot cards, the partner holding up the card can see the answer. If you make duplicate sets of cards, two pairs of students can use them at the same time. These cards can also be used for oral computation. After the child identifies the number represented, you could ask, "What would you have if I gave you three more? What would you have if I took four away?" and so forth.

1s-place addition cards: The four double-sided pages of 4.2a-d (pages 105-112) contain simple sums that involve adding only in the 1s place, without having to make a ten. Using their place-value mats and craft sticks, children can build the top number, then add sticks to represent the bottom number in order to determine the sum. They can also take turns drawing a card and solving the problem mentally. Because the answers are printed on the back, these cards are self-assessing if you copy both sides. If students need a prompt, encourage them to check their hands for the answer, rather than resorting to counting up. Children who become reliant on counting with their fingers have difficulty progressing beyond this strategy. Ignoring the 10s column, students can use "my two hands" to determine the total in the 1s column.

(For more advanced problems, they can use "my two hands" one column at a time.) The beauty of this method is that children never need to master facts over ten in order to solve any problem. (For a visual, see page 34.)

1s-place subtraction cards: The four double-sided pages labeled 4.3a-d (pages 113-120) are basic subtraction problems that do not require taking from the 10s place. Laminate and cut apart the cards, then use them as described in the previous paragraph for addition cards. (For a visual, see page 34.)

Make-a-ten addition cards: The sums in double-sided activities 4.4a-b (pages 121-124) require making a ten. It will be important to introduce these problems using the place-value mats. First the child would arrange sticks to represent the top number of the problem on the mat, then add the appropriate number of sticks for the bottom number of the problem. Remind the children, if necessary, that they need to "make a ten" by taking sticks from the 1s side to add to the new sticks in their hand, make a ten and place it in the 10s office. Then they can clear the mat, draw another card, and build the next problem.

SUBTRACTION, OR "TAKE FROM TEN"

Start with the children's place-value mats set up with 16 sticks (one 10 and six 1s) on them. (If you are continuing the lesson from the end of the story on page 22, there will already be 16 sticks on the place mats, or you could do a short addition story leading up to this sum.) As always, I use a story to introduce the new concept of "take from ten" when there are not enough sticks in the 1s office to subtract from there.

"Take from Ten" Story

You walk in the office door one morning and find that the air conditioning has broken and the office is hotter than hot! You choke and gasp and can hardly breathe! Just then five of the people start griping and complaining about how hot they are. You assure them you are going to fix the problem immediately, but they announce that they are leaving! They simply cannot work in that heat! You ask them to be patient, that you will work hard and fast to fix the problem, and soon the office will be nice and cool again. But they will not be consoled. They take their things and leave.

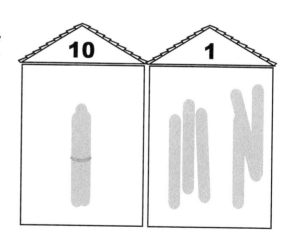

Processing: Ask the children where they would take the five people from. Guide them to look at the 1s side first. You want to build a habit of checking the 1s side first to determine whether there are enough people there from which to subtract the whole number.

Action: Having verified that there are enough people on the 1s side from which to take five, have the class do so together.

Synthesis: Now have children examine what is left. In this case, there will be one 10 and one 1 like this:

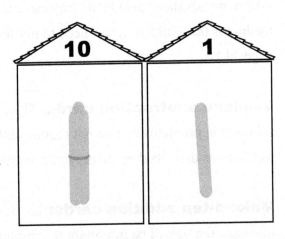

Story Continuation

On Monday morning when you come in the door, you smell a smell so horrible it almost makes you hurl! You recover, though, and stagger around the office to find the source of the stench. Then, just as everyone walks in to start work, you find a paper lunch bag sitting on the floor where Ms. Green left it last Friday when she left in a huff over the broken air conditioning. You look inside and find her spoiled tuna sandwich! Quickly you run the bag outside to the trash, but it's too late! The people who came in to start working are shouting, their faces are red, and they are holding their noses and waving their hands. Four of them insist they will not stay one minute longer, even though you found the source of the smell and assure them you are taking care of the problem immediately! No! They grab their things and leave. (You check behind them to make sure no one left his or her lunch behind!)

Processing: Ask the children, "Where did these four people leave from? Are there enough people on the 1s side for you to take four from? . . . No? . . . So where can we get four people? From the 10s side? Yes! That's right. We can 'take from ten!'"

Action: Guide the children to unbundle their set of ten and remove four people. Watch to see what they do with the remaining six sticks. If they hesitate and appear to be thinking about what to do, let them puzzle for a bit. If they automatically start to put the six sticks back on the 10s side, ask, "Is that a 10?" Give the children every opportunity to figure out on their own that they cannot put the six sticks on the 10s side anymore, and that they will need to be placed in the 1s office.

Synthesis: When you have worked through this point, the place-value mats will look like the one on the right. Talk with the students about what happened, and assess the number they have left.

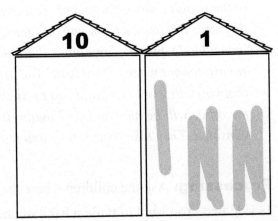

Continue with this type of storytelling until you see that the children automatically follow the correct procedure: they take from the 1s side if there are enough there to do so. If not, they move to the 10s. This is an important point to emphasize.

Math Center Games

Take-from-ten cards: 4.5a-c (pages 125-130) are similar to 4.3, except that they require unbundling a ten. Have students create the top number shown on the card on their place-value mats then subtract the bottom number. Doing so will require that they unbundle their ten, remove the correct number of sticks, and place the remaining sticks on the 1s side. (For a visual for "take from ten," see page 34.)

Pattern discovery: To promote pattern discovery, you might place in the math center only the cards in which nine is subtracted. Observe whether or not the students notice that when they unbundle their ten and take nine away, they always have one stick left over. Thus the answer to any "subtract nine" problem will always result in one more than the original number in the 1s office. Do not actively attempt to teach this pattern. Just note whether any of the students catch on to this rule through playing with only the minus nine cards. You should then proceed to providing only the minus eight cards, then both minus nine and minus eight cards. The goal of this game is student discovery rather than active teaching.

TAKING IT FURTHER

Working in small groups, lead the children to understand that the operations of "take from ten" and "make a ten" work exactly the same way for numbers larger than 20. Follow the steps in the "Story for Addition, or Make a Ten" and "Subtraction, or Take from Ten" sections, but this time, start with two or more bundles in the 10s office. Storytelling with more than one ten bundle will allow the students to recognize that the actions are identical for all computations using two digits, so once they have mastered this chapter, they will be able to do similar problems for numbers up to 99. I use the words "Beginning," "Action,"

Beginning = the initial number

Action = adding or subtracting

Ending = what is left after the action occurs

This language also makes a good connection to word problems.

and "Ending" to help children identify the three components of a problem and determine the correct procedure to use.

Beginning: Build a two-digit number.

Action: Add or subtract a one-digit number from the two-digit number. For example, you might start with three 10s and four 1s (see illustration). Present a story in which eight people want to leave (the action). First the children must check whether there are enough people in the 1s office to take the 8 from. No, there aren't. So what do they do? Unbundle a ten and take eight from there. Then the two leftover sticks go on the 1s side.

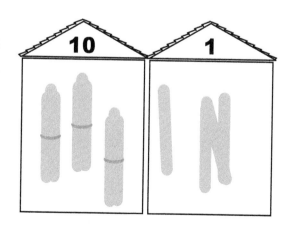

Ending: Assess what is left: "We used one 10 so that eight people could leave. That leaves us with two 10s in that office. We took away 8 from the 10 bundle and were left with two. These

-8

we put in the 1s side with the 4 sticks that were there already." Point out that the action still makes the shape of a backwards numeral 7: (1) Check the 1s side, (2) move to the 10s and unbundle, and (3) put any leftover sticks in the 1s office. Practice these types of problems until the children are fluent with procedures for addition and subtraction.

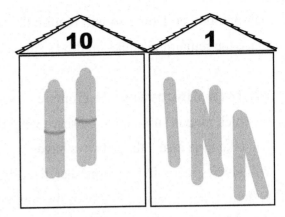

Subtracting Two-Digit Numbers

For a further challenge, set up the place-value mats with a number over 20 and lead students through subtracting a two-digit number, as described in the following example.

Beginning: Begin with two 10s and six 1s, as shown here.

Action: Present an action that does not require unbundling a ten; for example, ask students to subtract 13, or one 10 and three 1s. Guide them to notice that they are doing exactly what you said: taking one 10 from the 10s office and three 1s from the 1s office.

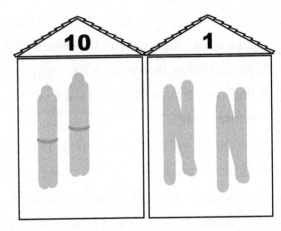

Ending: Analyze the pattern of sticks that remains after the action is completed (see illustration):

The value of proceeding all the way up to double-digit addition and subtraction is that students will not fear that computation will become harder as the numbers get bigger. If they gain mastery over the process for numbers below 20, they will have mastered the process for larger numbers as well. They can do it! This fact is empowering for all students but is especially critical for global learners.

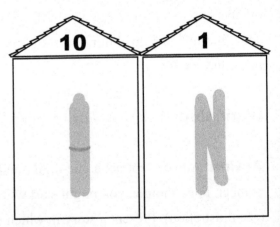

Math Center Games

Double-digit addition cards: Use 4.6a-b (pages 131-132) . The child draws a card (ex: + 12) and sets up that number on a place-value mat (bundle a 10 and have two 1s). Drawing a second card (ex: +19), the child will make the model of that second number and lay it on the table (bundle a 10 and have nine 1s). Many problems will require "make a ten" as the 1s side becomes too full. To help children manage the sticks without confusion, direct them to pick up the nine 1s, assess whether they need to make a ten out of the nine 1s in their hand and the two in the 1s office. To add 9 to 2, the child will take one from the 1s and put it with nine to make a 10. Now he will add it to the 10s side. All that is left of the 19 on the table is the ten which he can add to the 10s. What he now has is a 1, and three 10s: one from the 12, one from the 19, and the one he bundled. Then, leaving this set of sticks in place, the children draw another card, make the models, and add that number of sticks to those already on the mat.

Double-digit subtraction cards: The cards 4.7a-b (pages 133-134) work similarly to the addition cards, except the child should start with a very large number (such as 99) set up on the place-value mat. The child draws a card to subtract and starts by examining the 1s side. Are there enough people in that office to take away the number in the 1s place on the card? If so, the child simply takes the correct number of 1s from the 1s office and 10s from the 10s office. If there are not enough sticks on the 1s side, the child can pick up the number of 10s that will be subtracted plus one more to unbundle. Any leftover sticks from the unbundled 10 go on the 1s side. Allow plenty of practice with the addition and subtraction games.

Moneybags: Duplicate the pockets on 4.8 (page 135) and provide toy one- and ten-dollar bills. Label a series of items with prices in dollars only. (You could use empty food cartons, small toys, or simply pictures cut from catalogs and laminated.) The 1s pocket functions just like the 1s office, holding no more than nine dollars, and the 10s pocket parallels the 10s office. If you cut a slit partway along each pocket, the money can actually be placed in each pocket. One child pretends to be the customer and the other a sales clerk. The customer chooses a product to buy and checks the 1s pocket first to see if there is enough money there. If not, the customer will need to make change for a ten in order to pay, and then put the change in the 1s pocket. Introduce this game in a small group under your supervision until the children catch on. Then you can place the materials in the math center for independent use by pairs or small groups of children.

ASSESSMENTS

Assessment worksheets (4.9a-c) are provided on pages 136-138, and a progress report and whole-class record listing the target skills for chapters 3 and 4 are provided in Appendix B (pages 210-211). Instructions are provided on the worksheets: The child looks at the offices, adds or subtracts the number shown, and draws the result in the blank place-value card below. You will have gained a great deal of awareness about what each child knows from observing and working with small groups, but these materials will provide the final test of what children know. I would recommend testing more than once in order to gain an accurate reflection of what the children know. As always, I believe the primary purpose of assessment is to identify any areas that may require further teaching. Note that 4.9b involves adding and subtracting double-digit numbers. The students can solve these problems using the strategies they already know, but for children who are easily intimidated, you may wish to present only the single-digit problems.

Take from 10

Here is the picture of the problem 17 - 8 = 9.

You can't take the 8 from the 7 in the 1s place, so you have to break up the 10. See how it is cracking? The 8 we subtract is running away. The 2 left over from the broken 10 will join the 7 in the 1s place making the answer 2 + 7 = 9.

Make a 10

Here is the picture of the problem 6 + 4 = 10. You can use this for the problem 16 + 4 also. Basically, we are putting the 1s together to make a 10. Once we make the 10, the 1s are all used up and we have one 10 and zero 1s.

Subtracting 1s

Here is the picture of the problem 19 - 5 = 14. The 5 simply falls straight down out of the 9 in the 1s place and what is left is a 4 in the 1s place. We also can call this subtracting straight down.

Adding 1s

Here is the picture that shows what adding 1s looks like. The little numbers are piled up on top of each other and are simply going to fall straight down into the 1s place resulting in the 0 becoming a 6. The action is also called adding straight down.

5 A BIRD'S EYE VIEW

GOALS FOR THIS CHAPTER
1. To discover patterns in computation to 20
2. To gain a global understanding of computation to 20
3. To apply prior knowledge of computation to 10 to problems up to 20
4. To learn motions associated with subtraction and addition

This chapter presents the bird's-eye, or global, view of computation with numbers to 20 using a chart that organizes everything in a gorgeous whole! Seeing all the facts to 20 will satisfy those learners who need to know how much there is to learn and reveals patterns that will facilitate learning. Paired with the chart are motions that will aid recall: the motion for subtraction is a downward sweep, while the motion for addition is an upward sweep. The format of the chart connects this new information to students' previous learning of the number houses (*Right-Brained Addition & Subtraction*, Chapter 7). This chapter, in fact, begins with the 10s facts, providing a familiar point of reference. The children will see the new number facts as a continuation of what they already know.

Locate the chart entitled "Computation to 20: A Global View" (page 41) and enlarge it for class use. You also will want to make copies for individual student use (page 139.) If you laminate them, these pages can be used with dry-erase markers many times. The teaching procedure works through one section at a time, beginning with oral, small-group discussion and discovery, then moving to written practice. I recommend having the students sit as close to you and the chart as possible, to maximize focus and engagement. The blue part of the chart are all the problems that can be solved without disturbing the 10s. For example, in the column with 14 in the attic, you can subtract 14-14 by just subtracting straight down, 10-10 and 4-4. Likewise, you can subtract 14-13 by simply subtracting 10-10 and 4-3. This means that the children already know how to solve all the problems in blue simply by using their knowledge from 10th Street from the book *Right-Brained Addition & Subtraction*. For right-brained learners, showing them a chart like this is valuable. They will be able to see what they know and what they need to learn.

10		11		12		13		14		15		16		17		18		19		20	
0	10	0	11	0	12	0	13	0	14	0	15	0	16	0	17	0	18	0	19	0	20
1	9	1	10	1	11	1	12	1	13	1	14	1	15	1	16	1	17	1	18	1	19
2	8	2	9	2	10	2	11	2	12	2	13	2	14	2	15	2	16	2	17	2	18
3	7	3	8	3	9	3	10	3	11	3	12	3	13	3	14	3	15	3	16	3	17
4	6	4	7	4	8	4	9	4	10	4	11	4	12	4	13	4	14	4	15	4	16
5	5	5	6	5	7	5	8	5	9	5	10	5	11	5	12	5	13	5	14	5	15
				6	6	6	7	6	8	6	9	6	10	6	11	6	12	6	13	6	14
								7	7	7	8	7	9	7	10	7	11	7	12	7	13
												8	8	8	9	8	10	8	11	8	12
																9	9	9	10	9	11
																				10	10

STEP 1: PATTERN DISCOVERY

1. Display the chart and encourage the children to share what they notice. Allow plenty of time for them to reflect. Record their responses.

2. Focus on the 10s column, since the children are fluent with these problems. Point out that the ten is the attic number for this building. Explain to them that Stony Brook Village was growing so fast that the planning commission decided to build apartment buildings. Each building has an attic number, and each floor in the building has exactly that number of people. For example, in the 10s building, each floor has ten people in it. Have the children verify that this is indeed true.

3. Using the chart, make an arc with one hand across the two numbers in one equation and up to the ten in the attic (see illustration on page 38). Associate this motion of an upward curve, which looks like a C or backwards C, with addition.

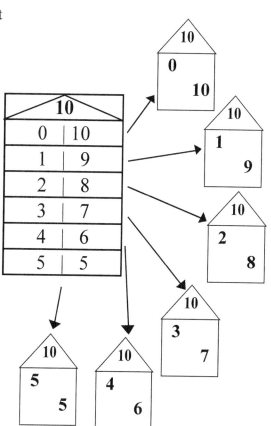

4. Have the children practice on their charts, reading an equation out loud while tracing the C or backwards C motion, moving upward from the lower numbers to the attic number.

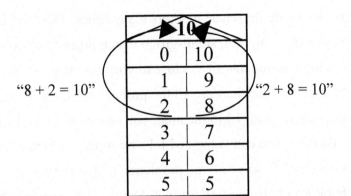

"8 + 2 = 10"   "2 + 8 = 10"

5. For an additional connection between the motion and the action of addition, model saying, "Come here" while gesturing in an upward sweep with your arm. Describe addition as being like more people coming and joining the group.

6. Now demonstrate that the motion for subtraction is the opposite of that for addition: It is still a C or backwards C, but this time the motion is down and away. You could tie this to a movement of waving people away, which would make the group smaller.

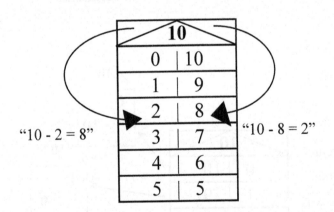

"10 - 2 = 8"   "10 - 8 = 2"

7. Have the children practice reading various equations while doing the addition or subtraction motions. I start with the familiar 10s facts at this stage, so children focus on the movement cues and then move to other columns.

STEP 2: WORKING WITH 1s

Draw attention to the fact that some of the floors in each building are enclosed in a box. Ask the children if they can guess why these floors are boxed. Give them time to brainstorm, and record their responses. Those floors in boxes are the problems that can be solved by adding or subtracting 1s (that is, without making or taking from ten). Point out that the students already know how to do all those problems. Choose any problem inside a box and show them that they can simply add or subtract the 1s and 10s without even touching the ten. (For example, 4 + 10 = 14 and 14 - 4 = 10.) The floors outside the boxes are the only problems that involve making a ten or taking from ten.

Practice

Oral practice: Give plenty of oral practice in solving problems inside the blue boxes. Look at a building, say number 13, and guide the children to recognize that each floor inside it is exactly like the threes houses, except that the 13 building has a ten by the attic number and also by the downstairs number. (That is, if the children ignore the numeral 1 in the 10s place, the problems become 0 + 3, 1 + 2, 2 + 1, and 3 + 0, all of which they know.)

Written practice: Give the children written practice with adding 1s and taking from 1s. (See activities 5.2a-c and 5.3a-c, pages 140-145.) Work for mastery, confidence, and fluency with computation.

ASSESSMENTS

Use the various pages of 5.2 and 5.3 to assess the students' knowledge of the material in this chapter. Use more than one sheet in making this determination. In addition, a reproducible whole-class record sheet and progress report for use with Chapters 5-7 are provided in Appendix B (pages 212-213).

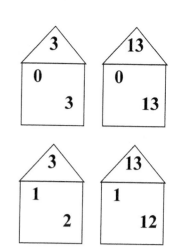

In the bottom row, 3 house repeats the one above with numbers switching floors.

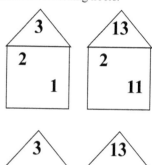

MAKE A 10

GOALS FOR THIS CHAPTER
1. To discover patterns in the global whole of math facts to 20
2. To determine how many equations there are for each number
3. To discover patterns that simplify problem solving
4. To gain fluency in solving problems that require "making a 10"

In this chapter, we will revisit the concept of "make a ten" using the global computation to 20 chart as a reference. (See page 41 for a copy.) We will be working only with addition in this chapter, and only with the problems outside the boxes on the computation chart on page 41.

Start out by having the children get out their charts that you photocopied for them earlier. Remind them that they already know all the equations inside the boxes! This is very exciting! Guide them to the realization that they have already mastered far more problems than they have left to master.

STEP 1: PATTERN DISCOVERY

Have the children examine the chart and discuss what they notice about the problems that lie outside the colored boxes. Allow plenty of time at this stage, as the children will be making discoveries and constructing

Computation to 20 - A Global View

20: 20/0, 19/1, 18/2, 17/3, 16/4, 15/5, 14/6, 13/7, 12/8, 11/9, 10/10

19: 19/0, 18/1, 17/2, 16/3, 15/4, 14/5, 13/6, 12/7, 11/8, 10/9

18: 18/0, 17/1, 16/2, 15/3, 14/4, 13/5, 12/6, 11/7, 10/8, 9/9

17: 17/0, 16/1, 15/2, 14/3, 13/4, 12/5, 11/6, 10/7, 9/8

16: 16/0, 15/1, 14/2, 13/3, 12/4, 11/5, 10/6, 9/7, 8/8

15: 15/0, 14/1, 13/2, 12/3, 11/4, 10/5, 9/6, 8/7

14: 14/0, 13/1, 12/2, 11/3, 10/4, 9/5, 8/6, 7/7

13: 13/0, 12/1, 11/2, 10/3, 9/4, 8/5, 7/6

12: 12/0, 11/1, 10/2, 9/3, 8/4, 7/5, 6/6

11: 11/0, 10/1, 9/2, 8/3, 7/4, 6/5

10: 10/0, 9/1, 8/2, 7/3, 6/4, 5/5

meaning for themselves. Here are some patterns to explore:

1. Every column has a "number + 9" equation right under its lowest colored box (see illustration). Run a finger across the problems inside the lavender boxes, so the children will notice this pattern. Let the children connect all the + 9 problems with a pencil, then count how many of these problems there are to learn. They might be gratified to note that there are only eight!

11		12		13		14		15		16		17		18	
0	11	0	12	0	13	0	14	0	15	0	16	0	17	0	18
1	10	1	11	1	12	1	13	1	14	1	15	1	16	1	17
2	9	2	10	2	11	2	12	2	13	2	14	2	15	2	16
3	8	3	9	3	10	3	11	3	12	3	13	3	14	3	15
4	7	4	8	4	9	4	10	4	11	4	12	4	13	4	14
5	6	5	7	5	8	5	9	5	10	5	11	5	12	5	13
		6	6	6	7	6	8	6	9	6	10	6	11	6	12
						7	7	7	8	7	9	7	10	7	11
										8	8	8	9	8	10
														9	9

2. Point out that they will not need to study the houses labeled 10 and 20 because they have already learned the 10s house, and the 20s house is just like it except that the 1s all make a ten and this makes the one in the 10s one digit larger. The 19 building is not included because all the problems just involve addition and subtraction of 1s.

11		12		13		14		15		16		17		18	
0	11	0	12	0	13	0	14	0	15	0	16	0	17	0	18
1	10	1	11	1	12	1	13	1	14	1	15	1	16	1	17
2	9	2	10	2	11	2	12	2	13	2	14	2	15	2	16
(3)	8	3	9	3	10	3	11	3	12	3	13	3	14	3	15
4	7	(4)	8	4	9	4	10	4	11	4	12	4	13	4	14
5	6	5	7	(5)	8	5	9	5	10	5	11	5	12	5	13
		6	6	6	7	(6)	8	6	9	6	10	6	11	6	12
						7	7	(7)	8	7	9	7	10	7	11
										(8)	8	8	9	8	10
														9	9

3. Now have the children locate the + 8 problems, right under the + 9 problems. Have them count how many there are. In a similar fashion, have them locate the "number + 7" and "number + 6" problems and count how many problems they will learn.

4. Identify that the numbers on the left side of each building also increase by one as you go down the chart (notice the green arrow pointing down), while on the right side of each building, the numbers increase going up (notice the blue arrow going up).

5. Now go back to the "number + 9" problems. Guide the children into noting a relationship between this problem and the attic number: The 1s place in the attic number is one number smaller than the number that is added to the nine. Have the children check all the buildings on the chart to verify that this pattern holds true.

1③		1④		1⑤	
0	13	0	14	0	15
1	12	1	13	1	14
2	11	2	12	2	13
3	10	3	11	3	12
④	9	4	10	4	11
5	8	⑤	9	5	10
6	7	6	8	⑥	9
		7	7	7	8

STEP 2: MAKE A TEN

Guide the students through oral practice with the "make a ten" problems in the houses, tying the equations to their previous practice with number houses and place-value mats.

Oral Practice

1. Starting with the 11 building, read the first equation under the box: "9 + 2 =11," making a C motion with your hand, as described in Chapter 5. Then do the same problem again, this time saying, "Nine 1s and two 1s is one 10 and one 1." Ask the children what happened to change the 1s into a mix of 10s and 1s. Guide them to recognize that because there were more than nine 1s in the house, it was necessary to "make a ten."

2. Ask, "How many 1s do we need to put with nine to make a ten?" If they hesitate, show them the 10's number house for 9 + 1 (see illustration). They can also look at their fingers, using the "my two hands" strategy to see how many they need to add.

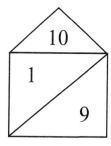

3. You might also want to relate what they just did to their practice with place-value mats and bundling to make a ten, as well as the Stony Brook number houses.

4. Write an addition problem in the traditional vertical format with the larger number on top (see illustration). **1.** They will take a 1 from the 4. **2.** Add the one to the 9 and make a ten. **3.** They can then write the 1 in the 10's place of the answer and write the 3 that was left from taking 1 from 4 and write it in the 1's place of the answer. Point out that the motions of the arrows in 1. - 2. are a backwards circle beginning at the middle right and continuing up and around to the left.

Written Practice

Give the children sheets of nines "make a ten" problems (6.1, page 146). Have them practice until everyone can solve the problems fluently and confidently. The children will benefit from picturing themselves holding the 9 in their hands as they pick up one from the smaller number to make a ten.

Finishing Up

Repeat the procedure described above for eights problems (6.2, page 146). Again, locate the problems on the chart, count how many there are, notice that the 1s place in the attic number is always two digits smaller than the number being added to the eight. To connect this pattern with prior learning, draw the $2 + 8 = 10$ house. Guide students to recognize that they always will need to take two 1s from the smaller number in order to make a ten out of the eight. Have them close their eyes and visualize doing this with sticks and place value mats. Once the students seem confident with the eights problems, present a sheet of mixed eights and nines "make a ten" problems (6.3, page 147).

Repeat the teaching process for the sevens equations (6.4, page 148). Next teach the sixes "make a ten" equations (6.5, page 148) and then present a mixed sheet of sixes and sevens "make a ten" problems (6.6, page 149). Next practice sixes, sevens, eights, and nines "make a ten" problems (6.7a-c, pages 150-152). Continue practicing until the children can confidently solve all the problems outside the boxes on the computation chart. The goal is to have the children super automatic on number pairs that make a ten so that when they see one number, they automatically know the partner: 1+9, 2+8, 3+7, 4+6, 5+5.

ASSESSMENTS

Use mixed-practice activities 6.7a-c (pages 150-152) for formal assessment. Look for any consistent patterns of problems missed to identify individual children's needs for further practice. If necessary, go back to practice with concrete materials with children who are really stuck. Also remind the children that if they have difficulty remembering how many they need to add to a number to make ten, they can use the "my two hands" strategy to refresh their memories. If children seem to hesitate with what to do when confronted with a problem, ask them to close their eyes and picture the action in the problem when they acted out that type of problem or when they used sticks and place value mats.

A reproducible whole-class record sheet and progress report for use with Chapters 5-7 are provided in Appendix B (pages 212-213).

7 TAKE FROM 10

GOALS FOR THIS CHAPTER

1. To discover patterns in the global whole of math facts to 20
2. To determine how many equations there are for each number
3. To discover patterns that simplify problem solving
4. To gain fluency in solving problems that require "making a 10"

In this chapter, we focus on subtraction with the problems on the global computation chart that lie outside the colored boxes. When I think back to the time when my third graders were most likely to turn pale, it was when they caught a glimpse of double-digit subtraction. There is something about this aspect of computation that strikes fear into the hearts of some children. For this reason, we are going to take the time to lay a good foundation for written double-digit subtraction. We do not want to lose any child at this point!

One thing we cannot do is simply to explain how to do subtraction in sequential steps. Expecting students to remember a series of steps they did not work through or discover for themselves will not work. Because some of the students will be unable to remember the steps, a verbal explanation does not save time in the long run. Instead we want to create an environment in which we can guide students to discover the action of the computation and to construct a procedure for themselves.

STEP 1: USE CONCRETE MATERIALS

Revisit the stories from Chapter 4 in which the Stony Brook offices were plagued with misfortune, causing some workers to leave (see pages 29-30). Provide place-value mats, craft sticks, and rubber bands for this review, and have the class physically work through the problems you choose. In your review, emphasize these points:

1. Subtraction involves people leaving.

2. Always check the 1s side first to see if there are enough people there from which to take the 1s who are leaving.

3. If there are not, you must take from ten by unbundling a ten and putting the remaining people in the 1s office. Remember you can't leave "loose" people in the 10's side.

TEACHING HINT

On a large card, print:

More (+) Means Make a Ten

Leave (-) Means Take from Ten

This will remind the children that with addition more come in, so they make a ten, whereas in subtraction some leave, so they take from ten.

STEP 2: INTRODUCE SYMBOLS

We will work through subtraction for one number at a time, so children have opportunities to discover the patterns that result from subtracting particular numbers. As with other steps in the Right-Brain Computation process, we will begin with pattern discovery, then use stories to provide a context for subtraction.

Pattern Discovery

1. Give each child a copy of the global computation chart you presented in Chapter 5 (5.1, page 139). Review how to use the chart for subtraction: The motion for subtraction is the opposite of that for addition: It is a C or backwards C starting at the attic number and moving down and over the two numbers that comprise the computation. You could tie this movement to waving people away, which would make the group smaller.

2. Refer to the global chart and locate the eight problems in the 11 to 18 columns that involve the number nine (the 1s right below the colored boxes). Recall from Chapter 5 that the problems below the boxes require using "make a ten" or "take from ten." Also recall that I do not focus on the 10, 19, and 20 columns because students have already learned those.

3. Write these problems on the board, leaving space between them. Make sure the children locate where these problems came from within the global chart:

11 △ -9	12 -9	13 -9	14 -9	15 -9	16 -9	17 -9	18 -9

4. See if the children comment on the sequence of attic numbers: the top numbers are all attic numbers. (Pattern: Each number increases by one as they progress to the right.)

5. Ask them to make predictions about the answers. Do they think a pattern will appear in the answers as well as in the attic numbers? Record their thoughts, but do not actively teach the answers at this point.

Oral Computation

Starting with the first problem (11 - 9) , ask the children, "What is happening in this story?" You want them to see the symbols but remember the action. For example:

Beginning: The 11 is the beginning of the story: Say, "Once upon a time there were 11 people hard at work," or present a similar scenario.

Action: Point to the - 9 row of the problem and ask, "What do you think happened in this second part of the story?" Let a volunteer invent a reason why nine people had to leave. (Maybe nine people broke out in a rash and had to go home!)

1.
$$\begin{array}{r} \cancel{1}1 \\ -\ \cancel{9} \end{array}$$

Ending: The ending of the story is the number of people who stayed at work.

1. Ask, "Could 9 people leave from the 1's office?" We always check the 1s office first. As soon as the class has identified that it will be necessary to "take from ten," cross out the numeral 1 in the 10s place and the numeral 9, saying, "Those nine people left from the 10's office. (See illustration 1.) How many were left over from the 10s we unbundled?" (One).

2.
$$\begin{array}{r} \cancel{1}1^{\,+\,1} \\ -\ \cancel{9} \end{array}$$

2. If the children reply correctly, ask where the one remaining person from the 10s office will go now. As they answer "to the 1s office," write + 1 by the right hand 1 of the 11. (See illustration 2.)

3.
$$\begin{array}{r} \cancel{1}1^{\,+\,1\,=\,2} \\ -\ \cancel{9} \\ \hline 2 \end{array}$$

3. Add the 1 + 1 in the 1s office and write the answer below the line. This is the end of that story. (See illustration 3.) Over time, and with a lot of practice, children might be able to just glance at the -9 and mentally add one to the 1s place and write the answer.

Point out how the procedure to solve the problem makes the shape of a Y. ("Y did they leave?" you could ask, as yet another mnemonic!)

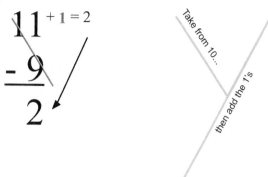

Move to the next problem (12 - 9) and repeat the process:

1. Ask, "What is the story?" (Twelve people were working, then nine left for some reason.)
2. Ask, "Can you take nine from the 1s place? ... No? So we take from ten and cross out the 1 in the 10s place and the 9."
3. Ask how many are left after the people left the 10s office. Write + 1 by the 2 in the 1s place.
4. Complete the problem by adding the 1s (that is, 2 + 1) and writing the answer.

Point out that the top part of the Y shape used to solve the problem is like a V. This shape will help students to remember how to check their work. Starting at the top left, draw the downward stroke of a V, saying "ten minus nine." Then complete the upward stroke, saying "is one." This body motion will help them remember to take the leftover 1s to the 1s side and add them to the 1s already there to arrive at the answer.

$$10 - 9 = 1 \quad \overset{+1=2}{11} \\ -9 \\ 2$$

STEP 3: SYNTHESIZE LEARNING

Complete the six other "subtract nine" problems in the same way. When all the problems have been solved on the board, ask whether the answers make a pattern:

- How do the answers compare to the attic numbers?
- Do you think that every time you subtract nine from a number, this pattern will appear? Why or why not? (See the illustration on page 50.)

Facts are taken from these 10s houses, as follows:

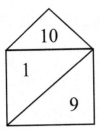 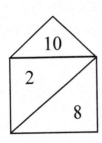 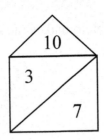 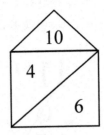 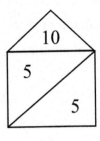

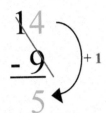  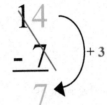 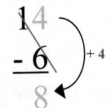 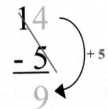

| Whenever you subtract 9, the answer will be one bigger than the 1s place in the starting number. | Whenever you subtract 8, the answer will be two bigger than the 1s place in the starting number. | Whenever you subtract 7, the answer will be three bigger than the 1s place in the starting number. | Whenever you subtract 6, the answer will be four bigger than the 1s place in the starting number. | Whenever you subtract 5, the answer will be five bigger than the 1s place in the starting number. |

STEP 4: PRACTICE WRITTEN COMPUTATION

See 7.1-7.9 (pages 153-166) to provide written subtraction practice.

1. Start with 7.1 (page 153) the nines "take from ten" problems. As students work, walk around and watch their progress, making notes about how each child is reacting to the work and solving the problems. Some children will move their pencils over each problem, mimicking the shapes you demonstrated, others will whisper the action to themselves, while others will simply look at the problem and quickly write the answer.

2. It is important at this stage to identify any child who is unable to do the work. If you realize that a child is struggling, work with him or her one-on-one:

•   Provide concrete materials to manipulate and have the child talk through the action of each problem while you ask prompting questions if necessary. Also ask him to close his eyes and picture the action.

•   Do not repeat the story/explanation you gave to the class. If that explanation had made sense to the child, he or she would be able to solve the problems. It might be that the child has to construct a process that fits his or her own particular way of understanding.

•   Encourage the child to talk about the problems until you identify the sticking point and can help him or her find a way to make sense of and recall the process.

3. Proceed through the remaining worksheets in this order:

•   7.2, eights "take from ten" problems (page 154)

•   7.3a, eights and nines "take from ten" problems (page 155)

- 7.3b, eights and nines "take from ten" problems (page 156)
- 7.4, sevens "take from ten" problems (page 157)
- 7.5a, sevens and eights "take from ten" problems (page 158)
- 7.5b, sevens and eights "take from ten" problems (page 159)
- 7.6, sevens - nines "take from ten" problems (page 160)
- 7.7, sixes "take from ten" problems (page 161)
- 7.8, twos - fives "take from ten" problems (page 162)

ASSESSMENTS

When a student seems confident in completing 7.1 to 7.8, use the four mixed subtraction problems worksheets (7.9a-d, pages 163-166) for assessment. Have the student complete more than one of these worksheets, so you can be sure of mastery. As always, your primary goal in assessment is not to assign a grade but to identify any lingering confusions or difficulties and provide remedial teaching as necessary until every child has achieved mastery. A reproducible whole-class record sheet and progress report for use with Chapters 5-7 are provided in Appendix B (pages 212-213).

8 TAKING STOCK

At this stage of the game, it will be important to take some time to determine how well your students have grasped the concepts you have presented. So before rushing on, let's allow all those busy brains to deepen their understanding and fluency. This chapter contains activities to build students' confidence and ease with computation, as well as assessments that should reveal exactly where they are in their understanding.

STEP 1: ORAL DISCUSSION

I would recommend beginning math class every day with oral discussion of math problems, until the children are able to evaluate a problem quickly and determine what action is involved. The main goal of this activity is that each child learns to determine independently what action occurs in all types of addition and subtraction problems. Here is the procedure for oral discussion:

1. Choose a variety of addition and subtraction problems, both ones that require making a ten or taking from ten and ones that add and take from 1s.

2. Group the class around a chalkboard or whiteboard, so everyone can see and participate easily.

3. Write the first problem on the board. I usually begin with an adding 1s problem (that is, one that does not require making a ten), such as that shown in the following illustration. Ask, "What are the three parts of this story?" Ask leading questions as necessary to guide the students in talking through the problem, as shown in the following example:

Problem involves adding 1s:

13 **Part 1, Beginning:** We have 13, or one 10 and three 1s.

+ 5 **Part 2, Action:** Five more 1s came in. We can just add 1s.

18 **Part 3, Ending:** Now we have one 10 and eight 1s.

---

4. Write a "take from 1s" problem (subtraction without taking from ten). To make the process as easy as possible, you may choose simply to rearrange the numbers from the first problem (see illustration). Again ask, "What are the three parts of the story?" The following is an example of the conclusions students would reach for this type of problem:

Problem involves subtracting 1s:

18 **Part 1, Beginning:** We have 18, or one 10 and eight 1s.

- 5 **Part 2, Action:** Five 1s leave. We can take them from the 1s.

13 **Part 3, Ending:** Now we have one 10 and three 1s.

---

5. Next present a "make a ten" problem and talk through the same steps of finding and describing the three story parts, as shown in the illustration:

Problem involves making a 10:

8 **Part 1, Beginning:** We have eight 1s and zero 10s.

+ 6 **Part 2, Action:** Six more 1s came in. We need to make a 10.

14 **Part 3, Ending:** Now we have one 10 and four 1s left over.

---

6. Finally, give an example of a "take from ten" problem, as shown in the following illustration:

Problem involves taking from 10:

13 **Part 1, Beginning:** We have 13, or one 10 and three 1s.

- 8 **Part 2, Action:** Eight 1s leave. We have to take from a 10.

5 **Part 3, Ending:** Now we have five 1s left over.

Now that you have presented a representative problem of each type, continue by presenting a random mix of problems.

STEP 2: ORAL ASSESSMENT

Make up flash cards representing all four types of story actions (or find coordinating flashcard sets at www.child1st.com). Have one child at a time come to your desk. Take five minutes with the child, asking him or her to describe the story for each problem. Make notes about your assessment of the child's understanding and competency. A whole-class record and individual progress report applicable to the four types of problems can be found in Appendix B (pages 212-213). Identify children who need more practice in a specific area, and meet with these children in a small group for more oral practice.

STEP 3: WRITTEN PRACTICE

The purpose of presenting written problems is to determine whether all the students can identify the right procedure to use for each problem when they have a sheet of mixed problems. Lead into the worksheets with a brief oral discussion of the different problem types, as described previously. Then give the children the following worksheets in the order specified:

• Mixed addition and subtraction in 1s place (8.1a-f, pages 167-172)
• Mixed addition and subtraction of nines (8.2a-b, pages 173-174)
• Mixed addition and subtraction of eights (8.3a-b, pages 175-176)
• Mixed addition and subtraction of sevens (8.4a-b, pages 177-178)
• Mixed addition and subtraction of sixes (8.5a-b, pages 179-180)
• Mixed addition and subtraction, sixes - nines (8.6a-c, pages 181-183)
• Mixed addition and subtraction, twos - fives (8.7a-b, pages 184-185)
• Mixed addition practice (8.8a-b, pages 186-187; these pages contain a mix of "make a ten" and "add 1s" problems)
• Mixed subtraction practice (8.9a-b, pages 188-189; these pages contain a mix of "take from ten" and "take from 1s" problems)
• Mixed addition and subtraction practice (8.10a-b, pages 190-191; these pages contain a mix of all four actions: add 1s, make a ten, take from 1s, and take from 10s.

STEP 4: WRITTEN ASSESSMENT

Once the students have completed this series of worksheets, you should be able to identify each child's level of understanding from the patterns of errors. Use these worksheets to identify each child's need for further teaching. Group children according to the concepts they need to review and provide small-group practice until they can solve the problems automatically without hesitating to decide what to do. Focus at this point on talking to the children about how they each see the action in the four kinds of problems. Let them develop their own way of solving the problems because what makes sense to them will be most useful!

When the children demonstrate mastery, proceed to multi-digit computation in Chapter 9. For the most efficient progress, it is critical not to move on to Chapter 9 until the children are confident in solving all the problems presented thus far.

9 TO THE TOP!

I will never forget arriving at this moment with my kindergarten group last year, seeing those little faces with their chubby cheeks, their eyes wide open when facing really big numbers. Children find something magical about numbers that are enormous. And when they find out that they can work with and solve problems using these numerical giants, their awe and pride know no bounds. In this chapter we will apply the same procedure students have already mastered to larger numbers.

Note: It is essential that all the children be comfortable with the work to this point before you introduce the activities in this chapter.

ADDING AND SUBTRACTING 1s

I begin by drawing the class together in a close group, then writing a double-digit addition problem such as this one on my board:

$$\begin{array}{r} 23 \\ +42 \\ \hline \end{array}$$

1. I ask the class whether they can solve this problem. When they say, "No," I reply, "I believe you can!" I let them look at the problem for a few seconds, just waiting to see what their comments will be.

2. Then, drawing a vertical line between the offices and a little roof, I ask, "How many 1s are there?"

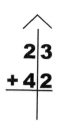

3. When they answer this question (five), I ask, "How many 10s are there?"

4. As the children answer, I write the numbers on the board.

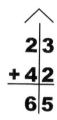

5. Next I choose another, similar problem that involves adding or subtracting 1s and 10s, but does not involve making or taking from ten. Ex: 16+53, 35+42, 72+15, 63+34, 81+17, etc.

6. I continue this exploration with the children until they realize that all they are doing is adding or subtracting numbers less than ten. I use the prompts, "Add straight down," (or "subtract straight down," for subtraction problems).

Oral Practice

1. Write multi-digit problems on the board and have the children discuss the answer in a group. Begin with a set of three-digit numbers that they can easily add column-by-column without making or taking from ten, such as the following:

$$\begin{array}{r} 2\,4\,1 \\ +3\,1\,4 \\ \hline 5\,5\,5 \end{array}$$

Then write a problem with four-digit numbers and repeat. Continue this way as long as the children are engaged and are enjoying this activity.

Written Practice

Copy and hand out any of the basic multi-digit addition and subtraction activities (9.1a-e, pages 192-196). I have included a series of these sheets that involve progressively larger numbers, but do not require making or taking from ten. Don't be surprised if you have some students who simply must find out how large the problem can get and still have this method work. Two of my students grappled with this process for a whole morning and were not satisfied until they had me turn the paper sideways and write problems that stretched the entire width of the page. When they found that this method worked even for problems that were about ten inches wide, they ran away to play. I like to imagine that they assumed at this point that there were no larger numbers, since we had used up the whole width of the paper!

DOUBLE-DIGIT MAKE A TEN

1. Show the children a problem that requires making a ten, such as the following. Discuss the "story" of this problem, using the procedure given in chapter 8.

2. Once the class has determined that the action requires them to make a ten, ask them how many 10s they will have once they make the new ten (two). Write the answer as they respond.

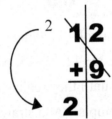

3. Next, ask them how many are left on the 1s side, now that they had to take some to make a ten.

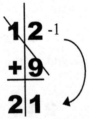

4. Guide them to reason that they took one from the two to make the ten, and now they have only one left. As the discussion proceeds, write - 1 (to symbolize the one taken from the two), then the answer at the bottom.

5. Recap the action: "We already had a ten on the 10s side, then we made another ten out of the nine plus one more. That gave us two 10s, and then we had only one left over in the 1s office. We wrote that one on the 1s side of the answer."

6. Continue to present similar problems until the children demonstrate that they understand what to do. I would suggest using problems such as these:

16	13	25	28	36	43
+9	+18	+15	+25	+28	+38

7. Talk through each problem in turn, modeling the steps on the board, as illustrated:

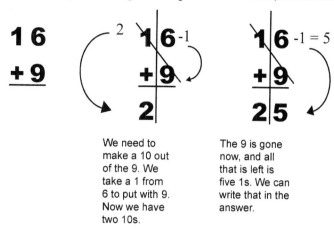

We need to make a 10 out of the 9. We take a 1 from 6 to put with 9. Now we have two 10s.

The 9 is gone now, and all that is left is five 1s. We can write that in the answer.

8. Continue in this fashion until the children are comfortable with the procedure and clearly understand what they are doing. If they understand the procedure, they will fall into problem-solving strategies that work well for them.

Work with Partners

Have the children work in pairs, taking turns solving double-digit problems. The child who is solving the problem talks through the process aloud. Meanwhile, the partner listens and monitors the first child's problem solving.

Written Practice

Locate the activities labeled "Double-digit make a ten" (9.2a-f, pages 197-199). Give the children half-sheets at first, explaining that when they finish that sheet, they are free to ask for more. Over time, the children will work more and more quickly, starting to do steps in their heads rather than on paper, until finally they can solve these problems entirely in their heads. This fluency results not from a lot of teaching, but from a lot of practice, allowing the children to arrive at steps that make the most sense to them.

SINGLE-DIGIT TAKE FROM TEN

Begin by talking through several problems as a class. Then, once the children are comfortable with the procedure, progress to written work.

Oral Introduction

1. Group the children around a white board or chalkboard. On the board, write a "take from ten" problem (see example).

$$\begin{array}{r} 1\,3 \\ -\,8 \\ \hline \end{array}$$

2. Talk through the problem-solving process by way of review:

• In this problem, the ten will disappear right away because we will need to take the eight from it.

• The two 1s that are left will go into the 1s office.

• Point out the Y-shaped movement as you first make a 10, and then bring 1s down to the answer.

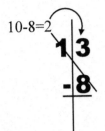

You can't take 8 from 3, so we unbundle a 10 to take the 8 from. Two 1s are left to put back on the 1s side.

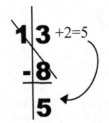

All we have now is three 1s and two more 1s. Add them and put the answer in the 1s place below.

3. Leaving this solved problem on the board, write **23 - 8**, in which the answer is exactly ten more.

4. Have the students establish that taking from ten is necessary, and then ask, "If I take a ten so that I can subtract the eight, how many 10s will be left?" (one) Write that in the answer space.

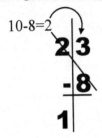

You can't take 8 from 3, so we unbundle a 10 to take the 8 from. We have one 10 left, so we write that in the answer. Two 1s are left to put back on the 1s side.

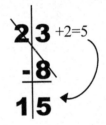

All we have now is three 1s and two more 1s. Add them and put the answer in the 1s place below.

5. Next, place the leftover 1s on the 1s side.

6. Now evaluate how many 1s there are and write the answer.

7. Discuss motions that the students might use as mnemonics for solving this type of problem. If they determine that they will need to take a ten, you will (1) slash through both the 10s place and the number being subtracted. (2) You will make a backwards C as you write the new 10s-place number at the top, write the one 10 in the answer, and (3) a backwards C as you add 3 + 2 on the 1s side, and then go to the bottom to write the answer.

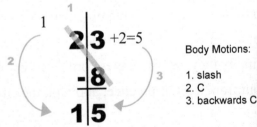

Body Motions:

1. slash
2. C
3. backwards C

8. At this point, write the problem 33-8 on the board and let the children predict what the answer will be. Solve this problem and compare the three answers. See if the children detect a pattern and can explain why this pattern is occurring. (The 1s place of the answer remains constant because the 1s in the problem do as well. The answer in the 10s place becomes one larger each time, just as the 10s place of the number being subtracted from does.)

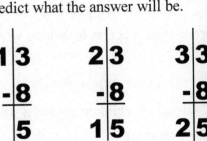

Synthesis

1. Write these five problems side-by-side on the board. **45 - 9, 45 - 8, 45 - 7, 45 - 6 and 45 - 5**. Solve these problems together and discuss the answers. Here are some tips students can use to self-check

their answers:

- When you take from ten, the answer in the 10s place will always be one smaller than the starting number.
- When you take from ten, the 1s side will always become larger because you put your leftovers there.

2. Construct the following chart, with the students helping to determine these facts:

- If you take away nine; the 1s side will increase by one   (10 - 9 = **1** and 9 + **1** = 10).
- If you take away eight, the 1s side will increase by two   (10 - 8 = **2** and 8 + **2** = 10).
- If you take away seven, the 1s side will increase by three   (10 - **3** = 7 and 7 + **3** = 10).
- If you take away six, the 1s side will increase by four   (10 - 6 = **4** and 6 + **4** = 10).

Continue for all the numbers to one. If needed, go back to place-value mats and sticks and have the children verify that these generalizations are indeed true.

3. Now write the following problem on the board:

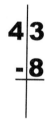

4. Have the children predict the answer, giving them plenty of time to think before you step in. If they are able to solve the problem mentally, proceed to the problem given in step 5. If they cannot solve mentally, ask questions to guide them through the steps:
- "Do we need to take from ten?" (yes)
- "How many 10s are left?" (two) Write 2 in the 10's place in the answer.
- "How many 1s are left from the 10 we unbundled?" (two) Write + 2 next to the 1s side.
- "How many 1s are there all together?" (five) Write the answer.

5. Next present the problem 53 - 8. Most likely the children will have the pattern down by now and will be able to arrive at the answer like this: "There will be one less 10, and two more 1s, so that makes four 10s and five 1s or 45. If they have difficulty, talk them through the steps as before. If any children hesitate over the answer, present the next problem in the series (63 - 8). Once the children are able to solve problems in this pattern confidently, write the problems that appear on the right. Talk through the process to arrive at answers as a class.

6 3	4 7	6 2	5 5	7 3
- 9	**- 8**	**- 5**	**- 6**	**- 7**
One 10 less, one more 1s or 54.	One 10 less, two more 1s or 39.	One 10 less, five more 1s or 57.	One 10 less, four more 1s or 49.	One 10 less, three more 1s or 66.

Written Practice

Locate 9.3a-b, labeled "Single-digit take from ten" (page 200). This page can be cut in half and presented one half at a time. Stay at this level for however long it takes for the children to become fluent with this type of problem. To be sure they have mastered the work to this point, give them a sheet of mixed problems (9.4a-b on page 201).

DOUBLE-DIGIT TAKE FROM TEN

These problems are much more complex, so introduce them slowly with oral practice at the beginning of each math lesson for several days. In actuality, the only added step is that 10s are being subtracted in two places: first you take a ten to make enough 1s, then you take however many 10s are needed for the digit in the 10s place. You may wish to stress this idea after you have worked a problem or two on the board.

Oral Introduction

1. Begin with the following problem:

$$\begin{array}{r} 2\,6 \\ -1\,9 \\ \hline \end{array}$$

Beginning:
Look at the 10's first. We have two 10s.

2. Ask the children to comment on what they see in the problem and how they might go about solving it. See whether any of the students recognize that they will need to take one of the 10s from which to subtract the nine, then use the other ten for subtracting the ten on the 10s side. Use arrows to illustrate this process.

2-2

Action:
How many 10's are we going to take from the 10s place? (2)

3. Once the 10s are gone, all that remains to be done is to put the leftover one (after taking nine from the ten) with the six and write the answer:

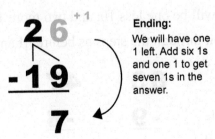

Ending:
We will have one 1 left. Add six 1s and one 1 to get seven 1s in the answer.

4. Evaluate the following problems (page 63), talking through the steps together in the same way:

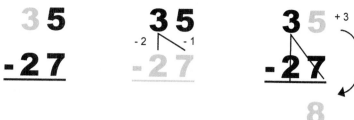

a. **35 - 27:** Take from ten because there aren't enough 1s.

b. How many 10s are we taking away? (Two full and one unbundled.) We write a small 2 for the two 10s the problem says to take away, and a 1 for the one 10 we unbundle. We add 2 + 1 to get the number of 10s we subtract. Are there any 10s left? (no).

c. Three leftover 1s go in the 1s side. Now add the 1s and write the answer on that side.

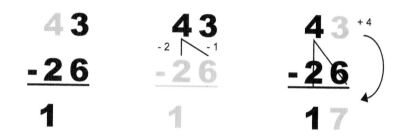

a. **43 - 26:** Take from ten because there aren't enough 1s.

b. How many 10s are we taking away? (Two full and one unbundled.) We write a small 2 for the two 10s the problem says to take away, and 1 for the one 10 we unbundle. We add 2 + 1 to get the number of 10s we subtract. Are there any 10s left? (yes) If so, write that number in the answer spot. (1)

c. Four leftover 1s go in the 1s side. Now add up the 1s and write the answer on that side. (7)

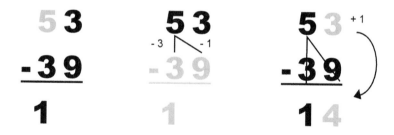

a. **53 - 39:** Take from ten because there aren't enough 1s.

b. How many 10s are we taking away? (Three full and one unbundled.) We write a small 3 for the three 10s the problem says to take away, and 1 for the one 10 we unbundle. We add 3 + 1 to get the number of 10s we subtract. Are there any 10s left? (yes). If so, write that number in the answer spot. (1)

c. Leftover 1s go in the 1s side. Now add up the 1s and write the answer on that side. (4)

5. Some children prefer to place the leftover 1s on the 1s side before dealing with the 10s side, which is fine. As teachers, our job is to be sure the children understand the problem-solving procedure, because once they do, *they will develop their own personal steps for solving the problems*. Many children at this point will just look at the 9 and automatically write the 1s answer as four. They will then be able to assess that four 10s are going to go away and see quickly that 5 - 4 = 1 for the 10s side.

Practice with Partners

Copy and cut apart 9.5a-b, entitled "Double-digit take from ten" (page 202). Present a half-worksheet at a time, and have children work in pairs to solve these problems. One child talks through the process of solving a problem while the other child observes and offers feedback. Then the partners can switch roles.

Written Practice

Let each child independently complete the same half-sheets (9.5a-b, page 202). As always, analyze the corrected papers to determine where each child needs more practice. Continue providing similar problems until the child appears to have mastered this problem type. Then provide at least two half-sheets of mixed problems for review and then for formal assessment use 9.6a-b, page 203.

AT THE TOP! MULTI-DIGIT COMPUTING

In the final section, we will present three-digit addition and subtraction problems. As before, begin by discussing the problems as a group, then move to small-group practice and written practice before finishing with formal assessment.

Oral Practice for Addition

1. Begin by presenting a problem such as the one on the right:

2. Evaluate the problem together. Don't write anything yet, just discuss what will happen in solving the problem:

$$\begin{array}{r} 237 \\ +389 \\ \hline \end{array}$$

- Establish that the only difference between three-digit computation and the problems the students already know how to solve is that there is one more column of numbers. The process is still identical.
- Look for places where making a ten will be necessary; namely, in the 1s column and the 10s column.
- Next, note that the students will have to take some 1s from the smaller number to make a ten. In the 1s column, the seven will become 7 - 1, and in the 10s column, the three will become 3 - 2. But then, remember that in making a ten, one more ten will show up in that column.

3. Now work through the computation step by step:

We will have to make a 10 and make a 100 in the first two columns. Let's go ahead and bundle, changing the numbers like in the illustration: adding a ten to the middle digit changes the 3 to a 4, and adding one 100 to the 100s office changes the 2 to a 3.

$$\begin{array}{r} {}^{3}\ {}^{4} \\ 237 \\ +389 \\ \hline \end{array}$$

Now let's write down what we took from the smaller number in each column to make our ten and 100. On the 1s side, we had to take one from the 7 to go with the 9. And in the 10s office (the middle numbers), we had to take 2 from the 4 to go with the 8.

$$\begin{array}{r} {}^{3}\ {}^{4-2} \\ 237\ {}^{-1} \\ +389 \\ \hline 6 \end{array}$$

Solve: 7 -1 = 6. Write the answer.

Next, 4 - 2 = 2. Write the answer.

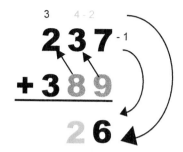
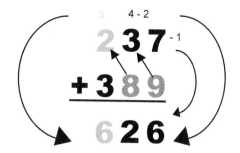

Now all we have to do is to add the last column. 3 + 3 = 6, which we write in the answer space.

If the children prefer to bundle first, then take from the smaller numbers, the steps are very similar.

1. Bundle the columns that need bundling, changing the top numbers to reflect this action.
2. Write the minus numbers by the top numbers (4 - 2 in the 10s column; 7 - 1 in the 1s column).
3. Solve each column, starting on the right. Write the answers in the answer field.
4. You're all done! Notice that the action forms a series of backslash lines, then arcs to the sides of the problem.

Work through problems aloud as a group until the children quickly see what they need to do.

Practice and Assessment

1. Have pairs of children work through problems together, as described in previous sections of this chapter.
2. Next have children solve the same or similar problems independently. Review their work to identify any misunderstandings requiring reteaching.
3. Finally, use the two half-worksheets labeled "Multi-digit make a ten" (9.7a-b, page 204) for formal assessment.

Oral Practice for Subtraction

1. As always, begin with oral group discussion of a problem such as the following:
2. Before writing anything, discuss the problem and what actions will be necessary in solving it: Starting on the right, ask if they can take the six from the three. No, so they will to go next door to the 10s office and unbundle a 10. Next, look at the 10s column. Ask whether they can take the five from the four. No! So they will need to go next door to the 100s office and and unbundle a 100!
3. Now that the children see what's in store, talk through solving the problem step by step, as shown in the example below.

$$743$$
$$-156$$

Evaluate: We cannot take 6 from 3 in the 1s column, nor 5 from 4 in the 10s column. We can take 1 from 7 in the 100s column. Draw lines through the numbers you are taking away from, like this, and write the new number above the old one.

In the 1s column, we can't take 6 from 3, so we will unbundle a 10. We no longer have four 10s. Write a 3 above the four.

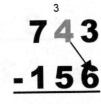

In the 10s column, we can't take a 5 from the 4, so we will unbundle a 100. We will only have six 100s left.

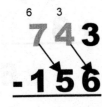

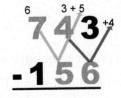

Next, put the leftovers by each top number. When you take 6 from 10, there will be 4 left over, so write + 4 by the 3. Notice the V shape you make when figuring out the leftover units.

Now let's solve! 3 + 4 is 7, and 3 + 5 is 8, so we can write those in the answer spaces.

Finally, we solve for 6 - 1 in the hundreds office and write that answer as well.

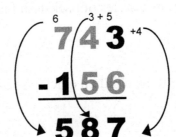

Some children may prefer to work one at a time—slashing through the four 10s and six 1s, then writing the four leftover 1s—before moving to the hundreds office. They would then slash through the seven and the five and write the leftover units (+ 5) in the 10s place.

As before, have the children practice with a partner first, then solve written problems independently. When your informal evaluation indicates that each child has mastered these problems, present the two halves of 9.8a-b (labeled "Multi-digit take from ten," page 205) for review and formal assessment. See Appendix B (pages 214-215) for a whole-class record and progress report.

CONCLUSION

Enjoy yourself. Encourage your students to decide now and then what math work they would like to do that day. I suspect you will find that they ask for specific areas in which they need more fluency. I would suggest keeping cubbies in the room with prepared sheets for each type of problem. On "free math day," students can simply choose whichever sheets they want to review. Allowing students to choose the math work they do is very empowering for them. They will delight in making decisions for themselves within the parameters you set for them.

APPENDIX A

ACTIVITIES & RESOURCES

2.1. Dot cards

1	2	3	4	5
6	7	8	9	10
11	12	13	14	15
16	17	18	19	20
21	22	23	24	25
26	27	28	29	30
31	32	33	34	35
36	37	38	39	40
41	42	43	44	45
46	47	48	49	50
51	52	53	54	55
56	57	58	59	60
61	62	63	64	65
66	67	68	69	70

1	2	3	4	5
6	7	8	9	10
11	12	13	14	15
16	17	18	19	20
21	22	23	24	25
26	27	28	29	30
31	32	33	34	35
36	37	38	39	40
41	42	43	44	45
46	47	48	49	50
51	52	53	54	55
56	57	58	59	60
61	62	63	64	65
66	67	68	69	70

2.3 Blank 5-frame chart to 20

2.4. Blank 5-frame chart

2.5a. 5-frame dot cards, page 1

Right-Brained Place Value, 2nd Edition, © 2013 Sarah Major, Child1st Publications, www.child1st.com.

2.5b. 5-frame dot cards, page 2

2.6a. Hollow dot cards, page 1

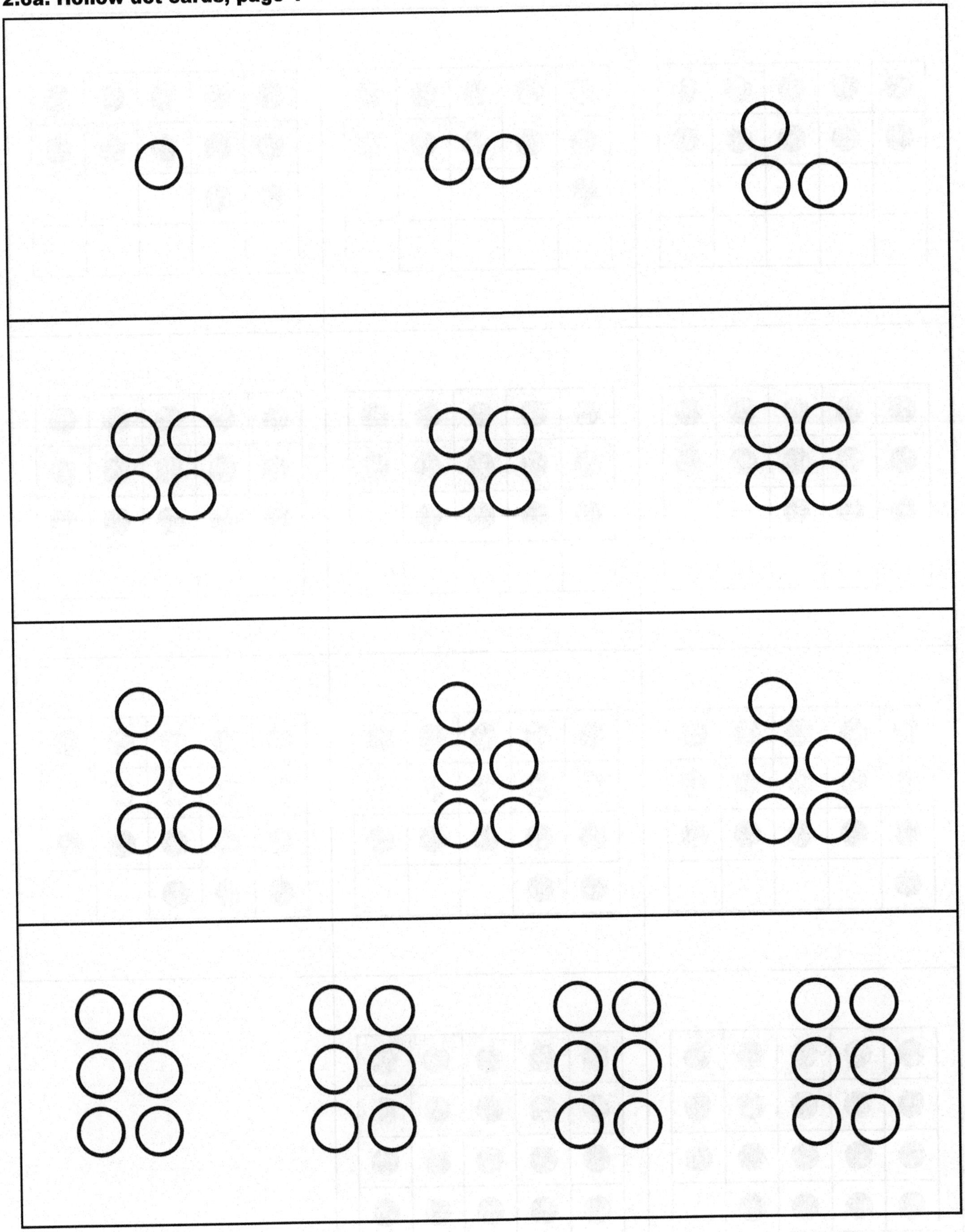

2.6b. Hollow dot cards, page 2

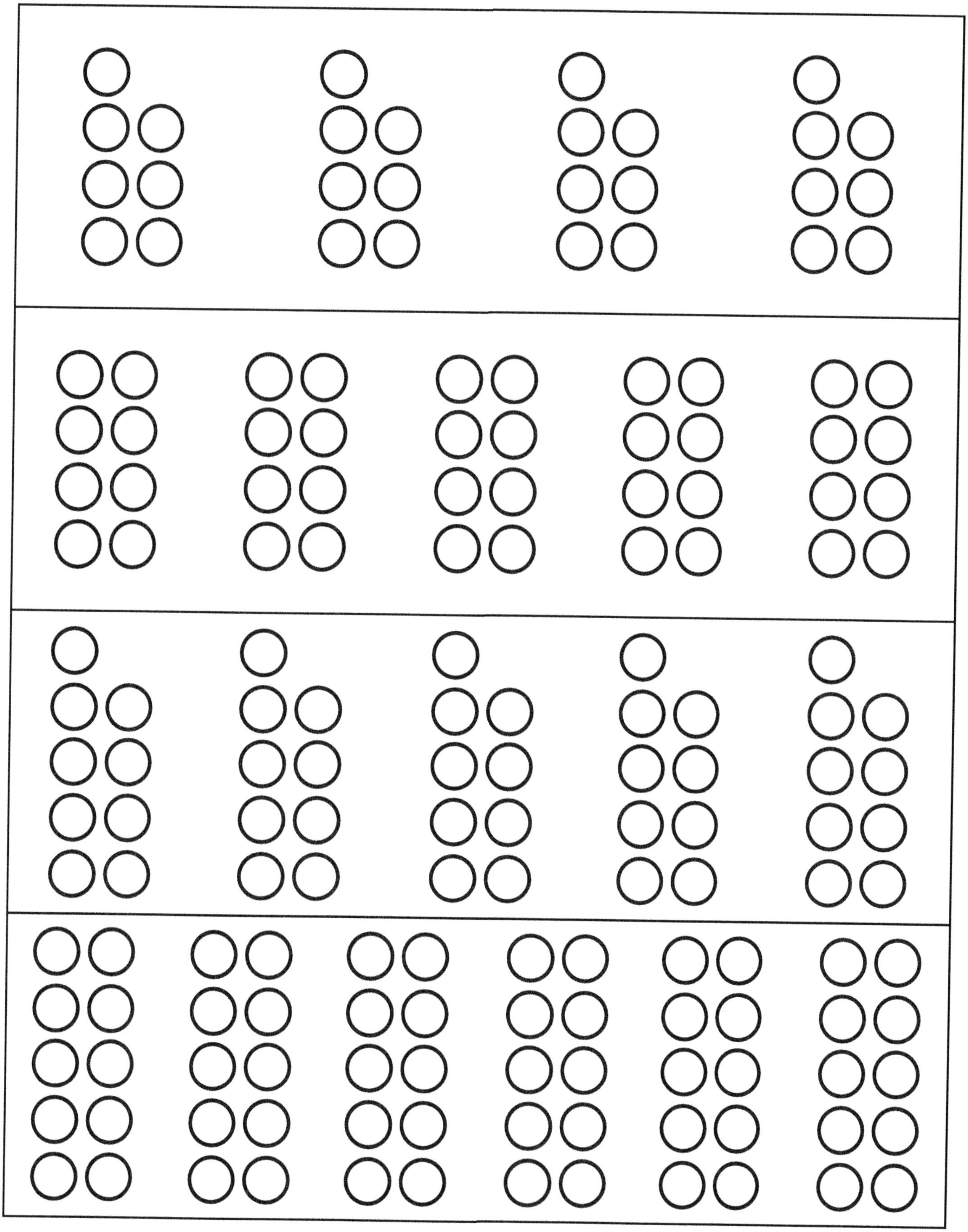

2.7a. Threes

Figure out which families can live in each house.

0 1 2 3

Name:_____

2.7b. Fours

Figure out which families can live in each house.

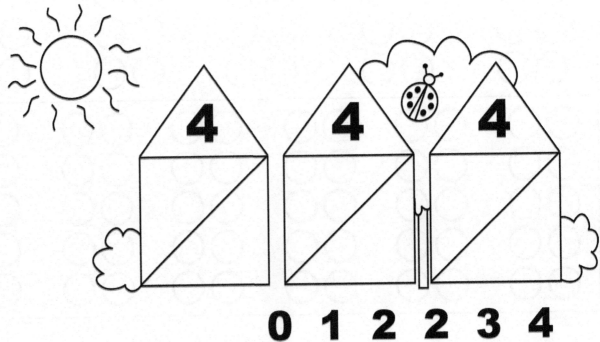

0 1 2 2 3 4

Name:_____

2.7c. Fives

Figure out which families can live in each house.

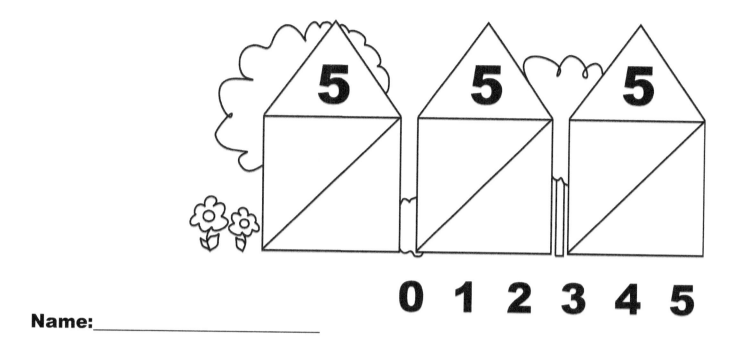

0   0   1   2   3   4   5

Name:_____

2.7d. Sixes

Figure out which families can live in each house.

0   0   1   2   3   3   4   5   6

Name:_____

Right-Brained Place Value, 2nd Edition, © 2013 Sarah Major, Child1st Publications, www.child1st.com.

Figure out which families can live in each house.

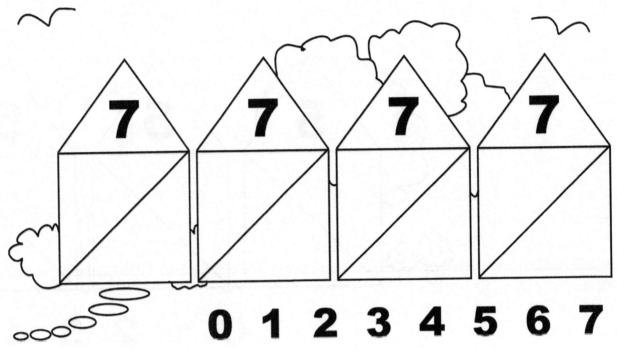

0 1 2 3 4 5 6 7

Name:_____

Figure out which families can live in each house.

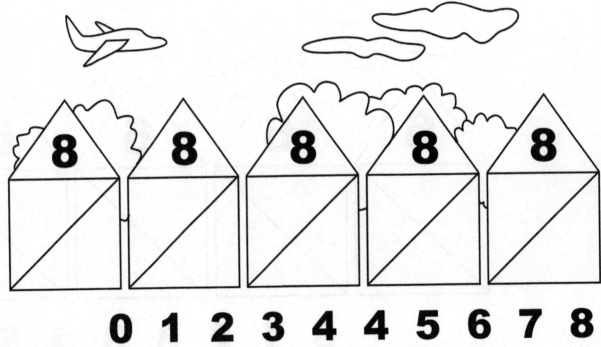

0 1 2 3 4 4 5 6 7 8

Name:_____

Right-Brained Place Value, 2nd Edition, © 2013 Sarah Major, Child1st Publications, www.child1st.com.

2.7g. Nines

Figure out which families can live in each house.

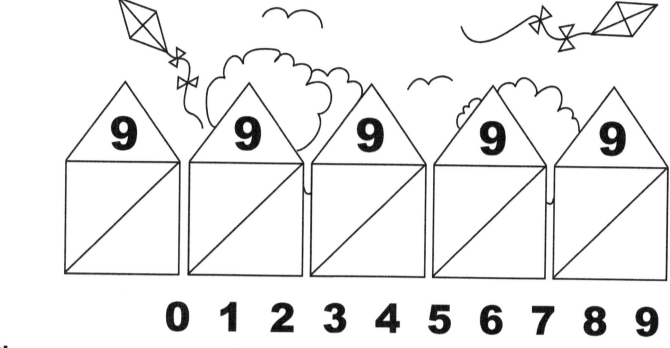

0 1 2 3 4 5 6 7 8 9

Name:_____

2.7h. 10s

Figure out which families can live in each house.

0 1 2 3 4 5 5 6 7 8 9 10

Name:_____

fold

fold

2.8a. Third Street practice houses

Right-Brained Place Value, 2nd Edition, © 2013 Sarah Major, Child1st Publications, www.child1st.com

fold

fold

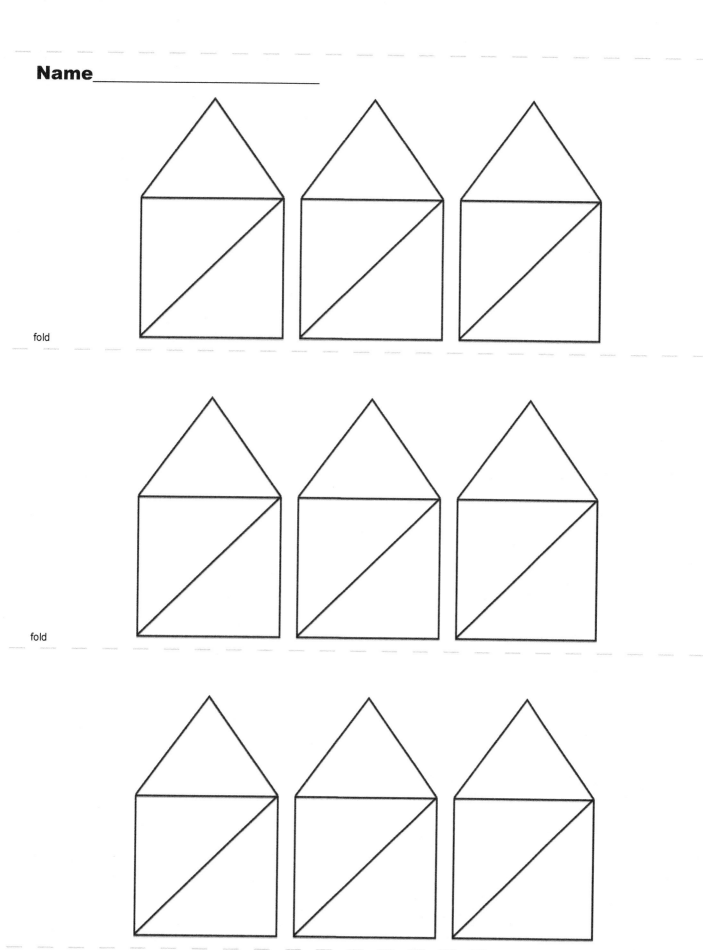

2.8b. Fourth and Fifth Street practice houses

Name_____

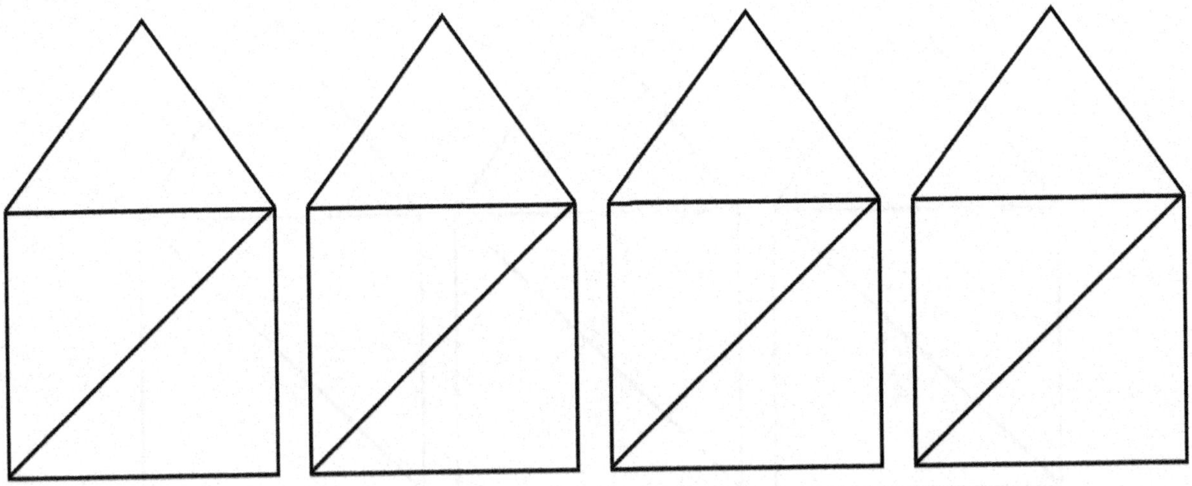

fold

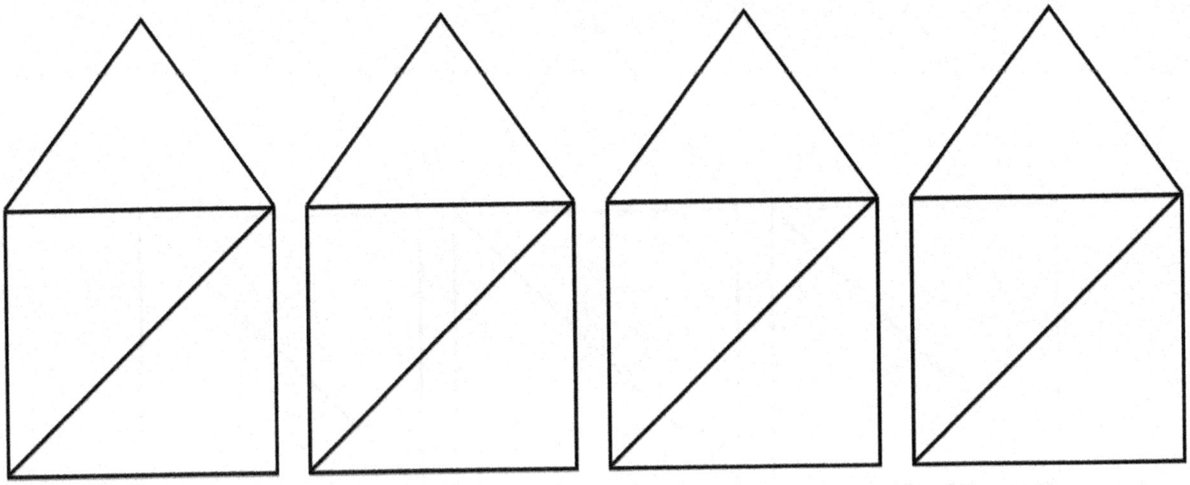

fold

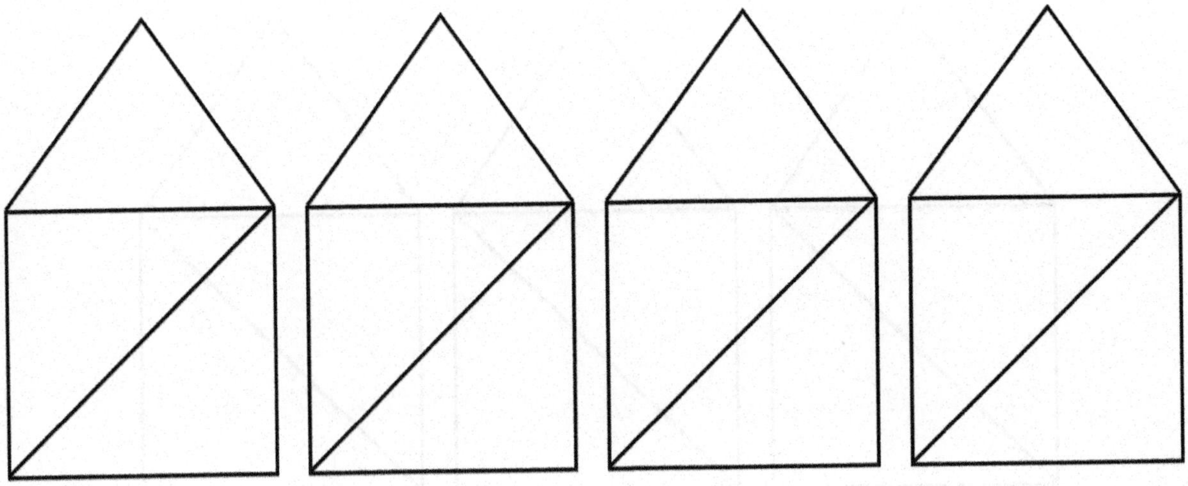

2.8c. Sixth and Seventh Street practice houses

Right-Brained Place Value, 2nd Edition, © 2013 Sarah Major, Child1st Publications, www.child1st.com.

Name_____

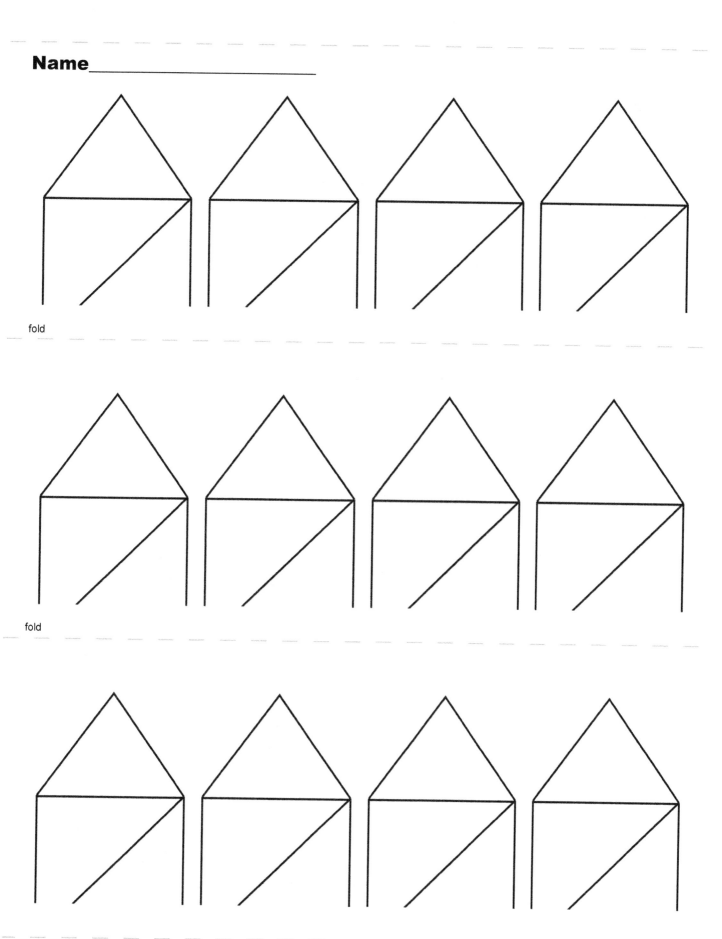

fold

fold

2.8d. Eighth and Ninth Street practice houses

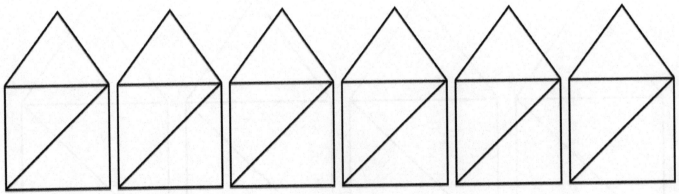

fold

fold

2.8e. Tenth Street practice houses

Name: _____

 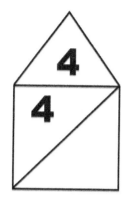 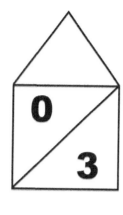  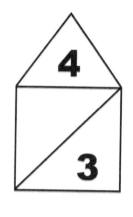

$$\begin{array}{ccccccc} 3 & 4 & 1 & 2 & 3 & 0 & 4 & 1 \\ -0 & -4 & +2 & +2 & -1 & +3 & -2 & +3 \end{array}$$

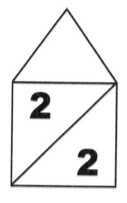 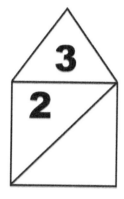   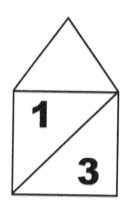

$$\begin{array}{ccccccc} 4 & 3 & 3 & 4 & 4 & 3 & 3 & 2 \\ -3 & -2 & +1 & +0 & -1 & +0 & -3 & +1 \end{array}$$

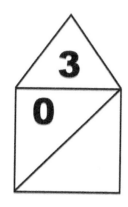 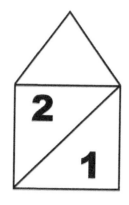 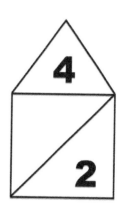 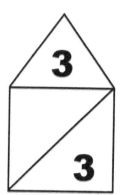

Name: _____

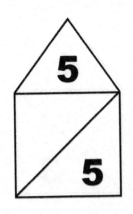   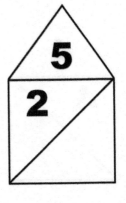 

6	5	0	3	5	4	5	1
- 5	- 3	+ 5	+ 3	- 1	+ 2	- 0	+ 4

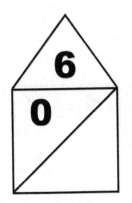  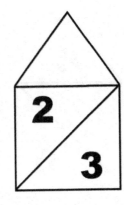  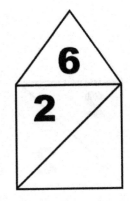

6	5	3	6	6	5	6	6
- 4	- 5	+ 2	+ 0	- 2	+ 1	- 3	- 1

  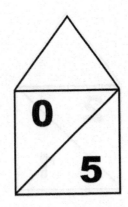  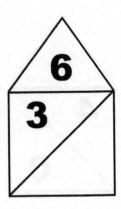

Right-Brained Place Value, 2nd Edition, © 2013 Sarah Major, Child1st Publications, www.child1st.com.

2.9c. Addition and subtraction practice - threes, fours, fives and sixes

Name: ______________________________

5 2
2 4
6 3
1 4
2 1
3 3
5 4
3 2
4 1
6 5
4 2
3 1
4 3
0 5
3 3
6 2
1 3
4 0
3 2
5 1
0 3
5 3
1 5
6 4
2 2

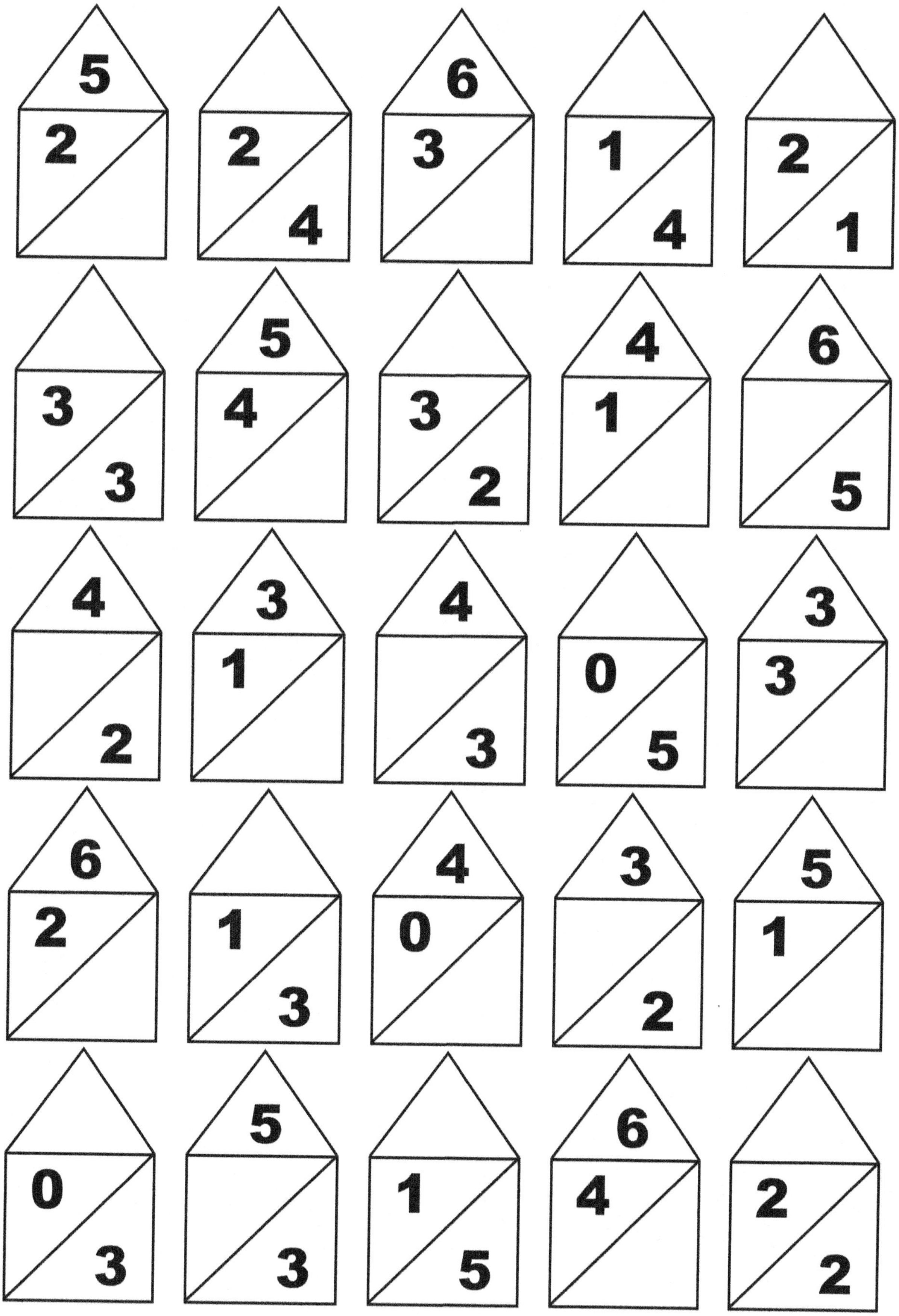

Name: _____

8	7	8	5	7	3	8	8
- 4	- 5	+ 0	+ 3	- 6	+ 4	- 2	- 8

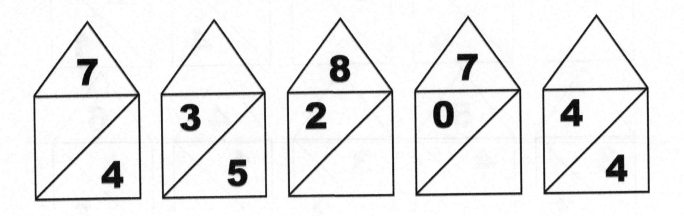

7	8	1	4	7	5	7	1
- 3	- 1	+ 6	+ 4	- 0	+ 2	- 1	+ 7

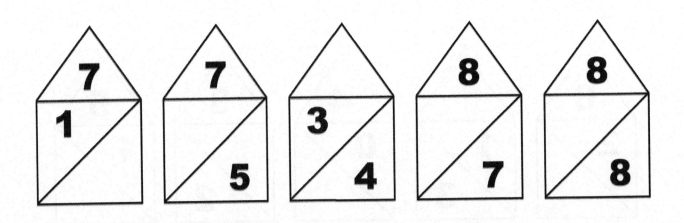

8	8	0	2	7	4	7	8
- 3	- 7	+ 7	+ 6	- 2	+ 3	- 4	- 6

2.9e. Addition and subtraction practice - nines and 10s

Name: _____

$$10 - 4 \qquad 9 - 5 \qquad 10 + 0 \qquad 6 + 3 \qquad 9 - 6 \qquad 1 + 8 \qquad 10 - 2 \qquad 9 - 9$$

$$10 - 7 \qquad 10 - 1 \qquad 7 + 2 \qquad 5 + 5 \qquad 10 - 0 \qquad 3 + 7 \qquad 9 - 1 \qquad 8 + 2$$

$$9 - 2 \qquad 10 - 8 \qquad 0 + 9 \qquad 4 + 6 \qquad 9 - 3 \qquad 4 + 5 \qquad 9 - 8 \qquad 10 - 5$$

Name: _____

Row 1: 9 / 3 | 2 / 8 | 10 / 4 | 2 / 7 | 1 / 6

Row 2: 5 / 5 | 9 / 4 | 1 / 8 | 8 / 3 | 10 / 7

Row 3: 8 / 7 | 7 / 3 | 8 / 6 | 0 / 9 | 7 / 2

Row 4: 10 / 1 | 4 / 4 | 8 / 0 | 7 / 4 | 9 / 1

Row 5: 0 / 7 | 9 / 5 | 1 / 9 | 10 / 3 | 3 / 5

3.1. Place-value transparency

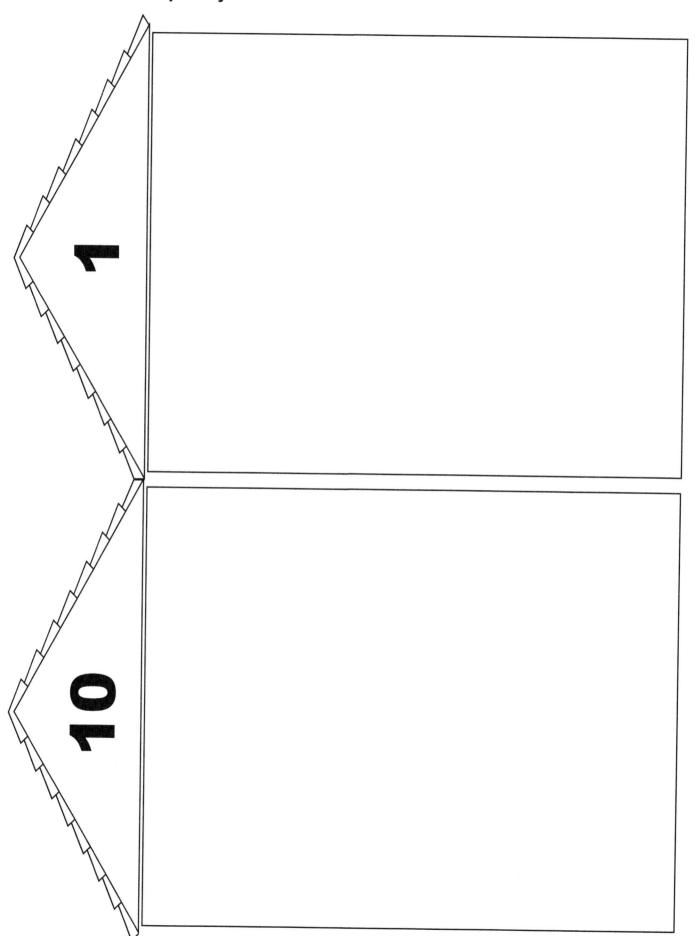

2 **people want to rent an office**	**2** **people want to rent an office**	**2** **people want to rent an office**
3 **people want to rent an office**	**3** **people want to rent an office**	**3** **people want to rent an office**
4 **people want to rent an office**	**4** **people want to rent an office**	**4** **people want to rent an office**
5 **people want to rent an office**	**5** **people want to rent an office**	**5** **people want to rent an office**

6 **people want to** **rent an office**	**6** **people want to** **rent an office**	**6** **people want to** **rent an office**
7 **people want to** **rent an office**	**7** **people want to** **rent an office**	**7** **people want to** **rent an office**
8 **people want to** **rent an office**	**8** **people want to** **rent an office**	**8** **people want to** **rent an office**
9 **people want to** **rent an office**	**9** **people want to** **rent an office**	**9** **people want to** **rent an office**

3.3b. Place-value mat 10

3.4. Blank place-value cards

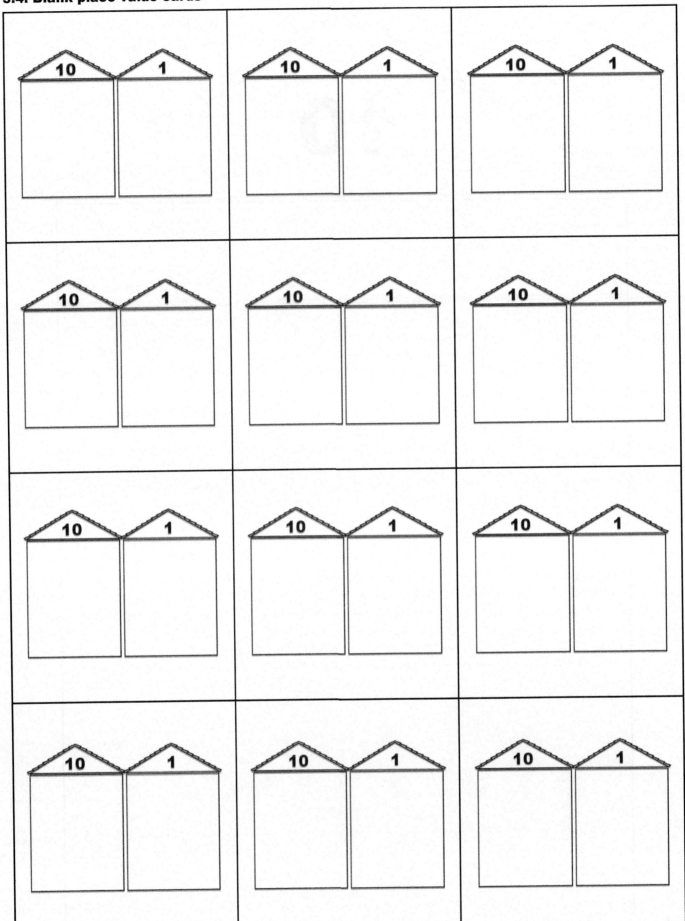

Right-Brained Place Value, 2nd Edition, © 2013 Sarah Major, Child1st Publications, www.child1st.com.

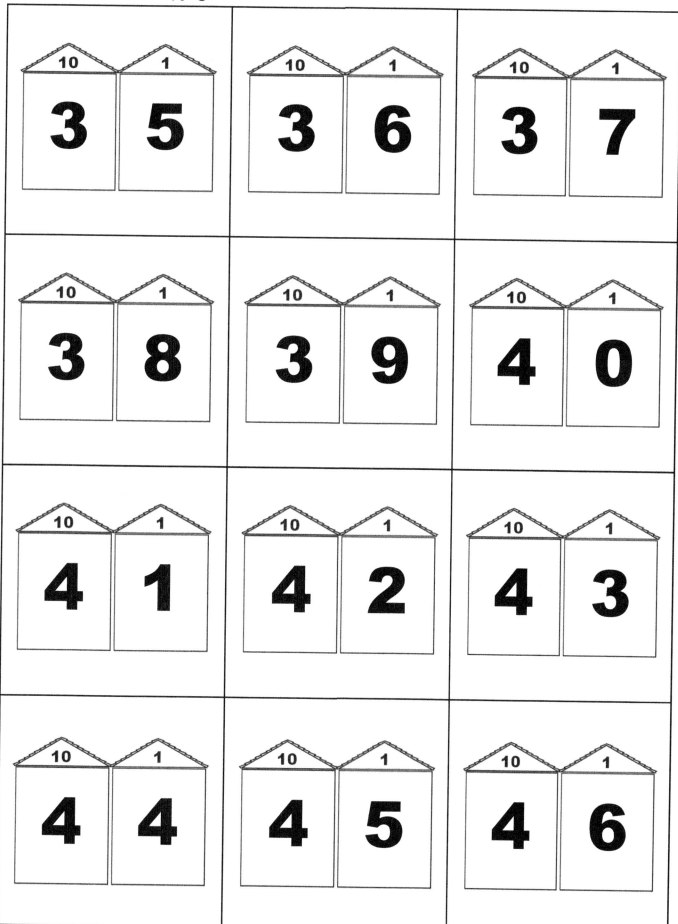

Name: _____

When the teacher says a number, write it in the 10 and 1 offices. For example, if the teacher said "twenty-three," you would write:

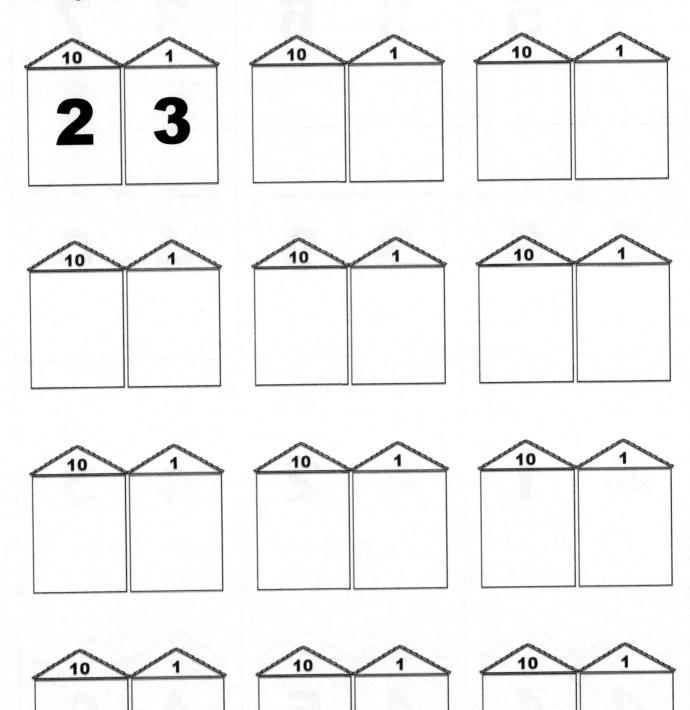

Right-Brained Place Value, 2nd Edition, © 2013 Sarah Major, Child1st Publications, www.child1st.com

Name: _____

Write the number shown in each office on the line underneath it.

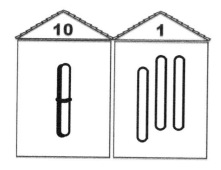

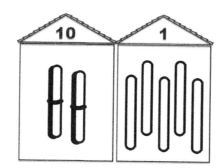

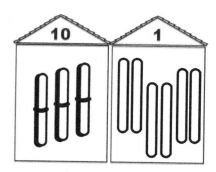

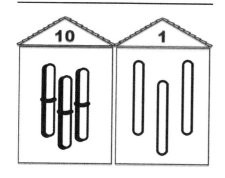

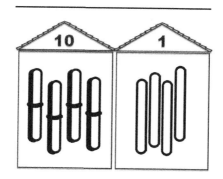

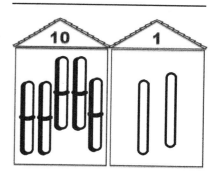

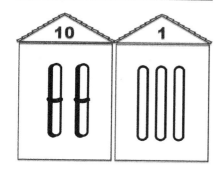

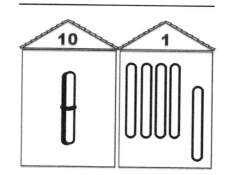

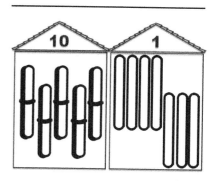

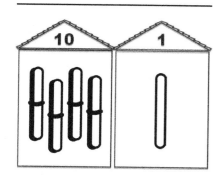

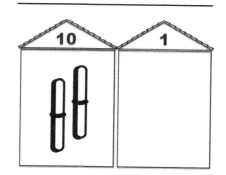

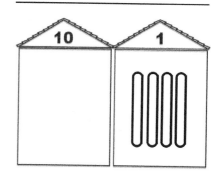

Name: _____

Draw a picture to show each number. The first one is done for you.

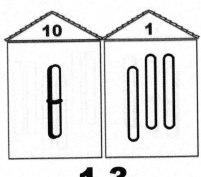 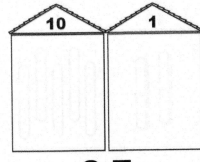 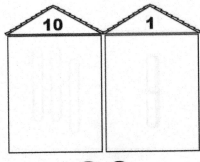

1 3   3 7   2 8

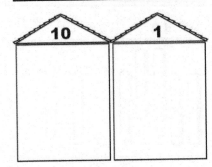 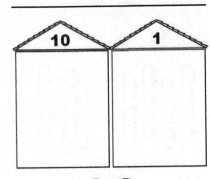 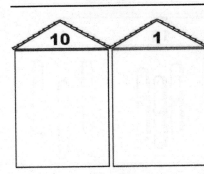

5 1   6 2   7 3

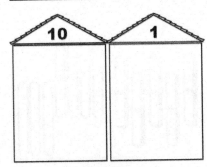 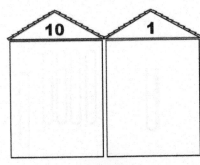 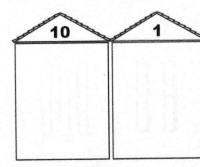

4 6   3 5   8 1

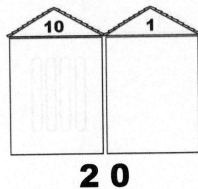 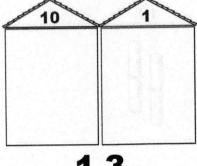 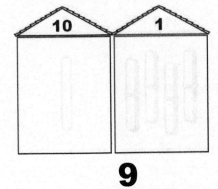

2 0   1 3   9

Right-Brained Place Value, 2nd Edition, © 2013 Sarah Major, Child1st Publications, www.child1st.com.

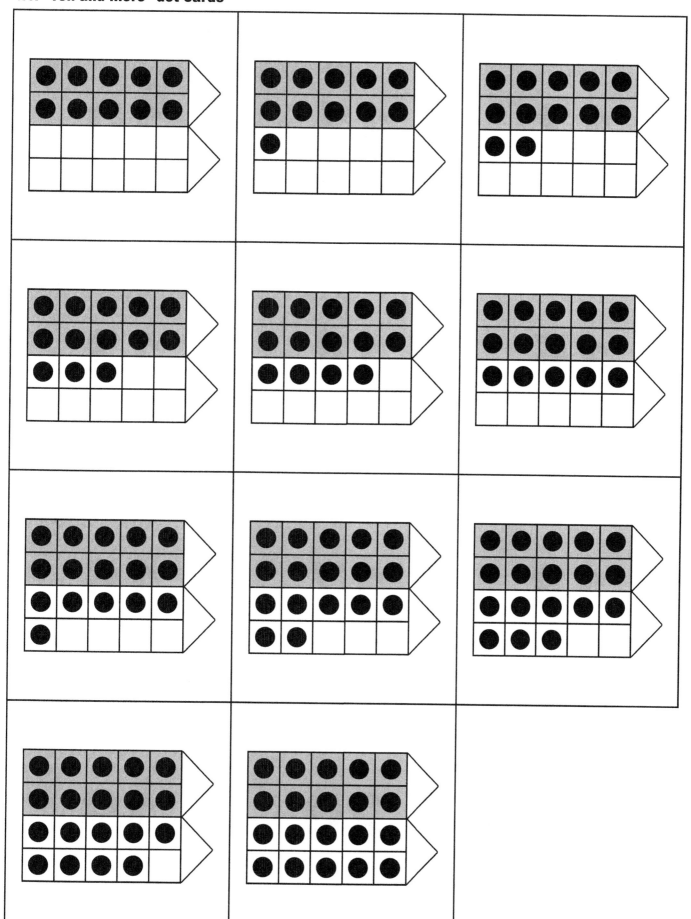

12 11 10

15 14 13

18 17 16

20 19

$\begin{array}{r} 10 \\ +\ 1 \\ \hline \end{array}$	$\begin{array}{r} 10 \\ +\ 2 \\ \hline \end{array}$	$\begin{array}{r} 10 \\ +\ 3 \\ \hline \end{array}$
$\begin{array}{r} 10 \\ +\ 4 \\ \hline \end{array}$	$\begin{array}{r} 10 \\ +\ 5 \\ \hline \end{array}$	$\begin{array}{r} 10 \\ +\ 6 \\ \hline \end{array}$
$\begin{array}{r} 10 \\ +\ 7 \\ \hline \end{array}$	$\begin{array}{r} 10 \\ +\ 8 \\ \hline \end{array}$	$\begin{array}{r} 10 \\ +\ 9 \\ \hline \end{array}$
$\begin{array}{r} 11 \\ +\ 1 \\ \hline \end{array}$	$\begin{array}{r} 11 \\ +\ 2 \\ \hline \end{array}$	$\begin{array}{r} 11 \\ +\ 3 \\ \hline \end{array}$

13	12	11
16	15	14
19	18	17
14	13	12

11 + 4	11 + 5	11 + 6
11 + 7	11 + 8	12 + 1
12 + 2	12 + 3	12 + 4
12 + 5	12 + 6	12 + 7

17	**16**	**15**
13	**19**	**18**
16	**15**	**14**
19	**18**	**17**

Right-Brained Place Value, 2nd Edition, © 2013 Sarah Major, Child1st Publications, www.child1st.com

13 + 1	13 + 2	13 + 3
13 + 4	13 + 5	13 + 6
14 + 1	14 + 2	14 + 3
14 + 4	14 + 5	15 + 1

16	15	14
19	18	17
17	16	15
16	19	18

15 + 2	15 + 3	15 + 4
16 + 1	16 + 2	16 + 3
17 + 1	17 + 2	18 + 1

19	**18**	**17**
19	**18**	**17**
19	**19**	**18**

19 - 9	19 - 8	19 - 7
19 - 6	19 - 5	19 - 4
19 - 3	19 - 2	19 - 1
18 - 8	18 - 7	18 - 6

12	11	10
15	14	13
18	17	16
12	11	10

18 - 5	18 - 4	18 - 3
18 - 2	18 - 1	17 - 7
17 - 6	17 - 5	17 - 4
17 - 3	17 - 2	17 - 1

15	**14**	**13**
10	**17**	**16**
13	**12**	**11**
16	**15**	**14**

16 **- 6**	16 **- 5**	16 **- 4**
16 **- 3**	16 **- 2**	16 **- 1**
15 **- 5**	15 **- 4**	15 **- 3**
15 **- 2**	15 **- 1**	14 **- 4**

12	11	10
15	14	13
12	11	10
10	14	13

14 - 3	14 - 2	14 - 1
13 - 3	13 - 2	13 - 1
12 - 2	12 - 1	11 - 1

13	**12**	**11**
12	**11**	**10**
10	**11**	**10**

9 + 9	9 + 8	9 + 7
9 + 6	9 + 5	9 + 4
9 + 3	9 + 2	9 + 1
8 + 8	8 + 7	8 + 6

16	**17**	**18**
13	**14**	**15**
10	**11**	**12**
14	**15**	**16**

$\begin{array}{r} 8 \\ + 5 \\ \hline \end{array}$	$\begin{array}{r} 8 \\ + 4 \\ \hline \end{array}$	$\begin{array}{r} 8 \\ + 3 \\ \hline \end{array}$
$\begin{array}{r} 8 \\ + 2 \\ \hline \end{array}$	$\begin{array}{r} 7 \\ + 7 \\ \hline \end{array}$	$\begin{array}{r} 7 \\ + 6 \\ \hline \end{array}$
$\begin{array}{r} 7 \\ + 5 \\ \hline \end{array}$	$\begin{array}{r} 7 \\ + 4 \\ \hline \end{array}$	$\begin{array}{r} 7 \\ + 3 \\ \hline \end{array}$
$\begin{array}{r} 6 \\ + 6 \\ \hline \end{array}$	$\begin{array}{r} 6 \\ + 5 \\ \hline \end{array}$	$\begin{array}{r} 6 \\ + 4 \\ \hline \end{array}$

11	12	13
13	14	10
10	11	12
10	11	12

Right-Brained Place Value, 2nd Edition, © 2013 Sarah Major, Child1st Publications, *www.child1st.com*

$\begin{array}{r} 11 \\ -\ 9 \\ \hline \end{array}$	$\begin{array}{r} 11 \\ -\ 8 \\ \hline \end{array}$	$\begin{array}{r} 11 \\ -\ 7 \\ \hline \end{array}$
$\begin{array}{r} 11 \\ -\ 6 \\ \hline \end{array}$	$\begin{array}{r} 11 \\ -\ 5 \\ \hline \end{array}$	$\begin{array}{r} 11 \\ -\ 4 \\ \hline \end{array}$
$\begin{array}{r} 11 \\ -\ 3 \\ \hline \end{array}$	$\begin{array}{r} 11 \\ -\ 2 \\ \hline \end{array}$	$\begin{array}{r} 12 \\ -\ 9 \\ \hline \end{array}$
$\begin{array}{r} 12 \\ -\ 8 \\ \hline \end{array}$	$\begin{array}{r} 12 \\ -\ 7 \\ \hline \end{array}$	$\begin{array}{r} 12 \\ -\ 6 \\ \hline \end{array}$

4	**3**	**2**
7	**6**	**5**
3	**9**	**8**
6	**5**	**4**

12 - 5	12 - 4	12 - 3
13 - 9	13 - 8	13 - 7
13 - 6	13 - 5	13 - 4
14 - 9	14 - 8	14 - 7

9   8   7

6   5   4

9   8   7

7   6   5

14 − 6	14 − 5	15 − 9
15 − 8	15 − 7	15 − 6
16 − 9	16 − 8	16 − 7
17 − 9	17 − 8	18 − 9

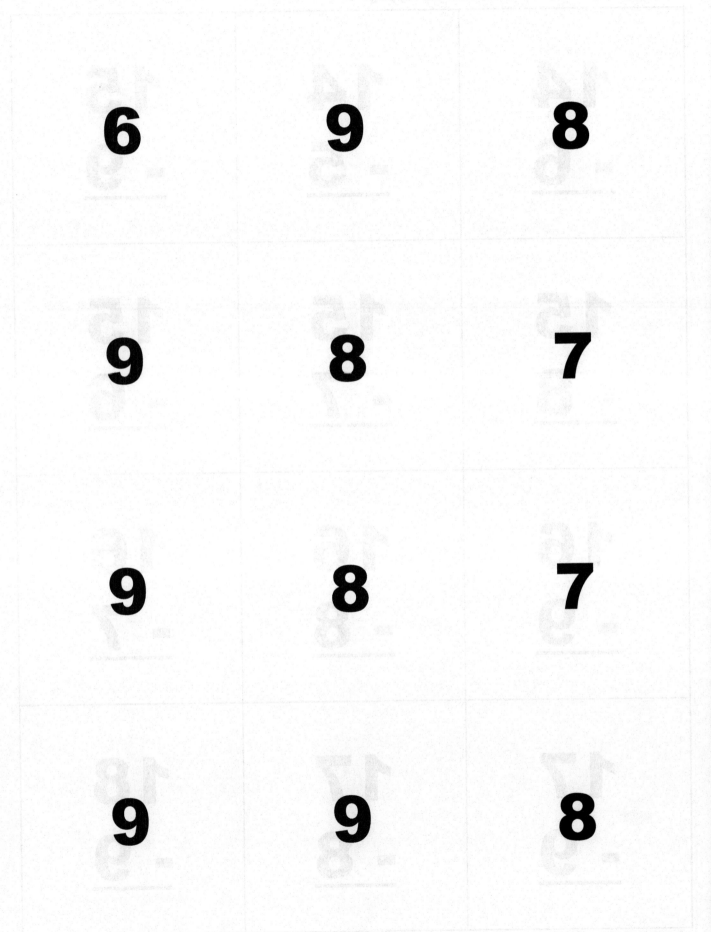

6	9	8
9	8	7
9	8	7
9	9	8

+ 10	+ 11	+ 12
+ 13	+ 14	+ 15
+ 16	+ 17	+ 18
+ 19	+ 20	+ 21

+ 22	+ 23	+ 24
+ 25	+ 26	+ 27
+ 28	+ 29	+ 30
+ 31	+ 32	+ 33

- 10	- 11	- 12
- 13	- 14	- 15
- 16	- 17	- 18
- 19	- 20	- 21

- 22	- 23	- 24
- 25	- 26	- 27
- 28	- 29	- 30
- 31	- 32	- 33

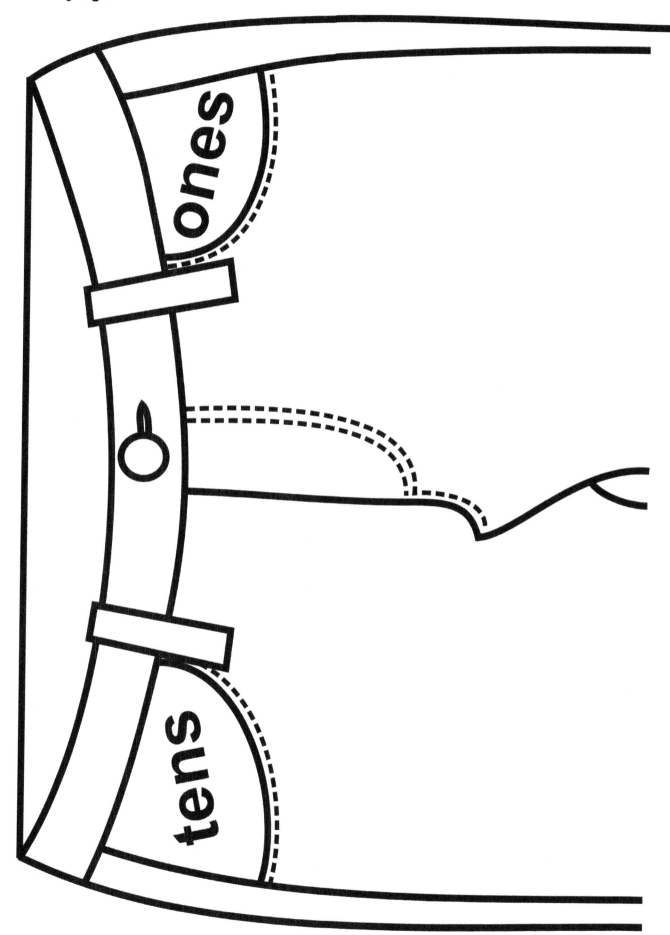

Name: _____

+ 4   **+ 8**   **- 3**

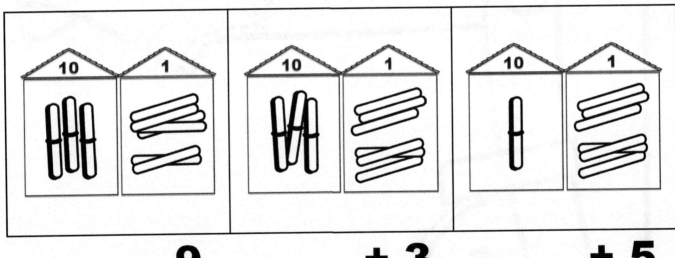

- 9   **+ 3**   **+ 5**

Name: _____

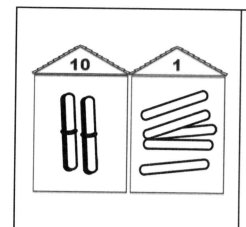

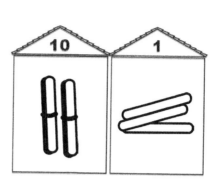

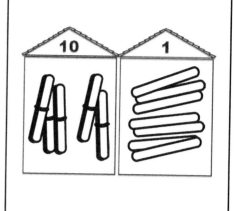

+ 1 2

+ 2 6

- 3 6

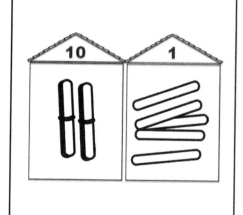

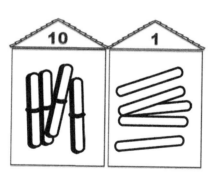

+ 4 3

- 4 1

- 2 3

Name: _____

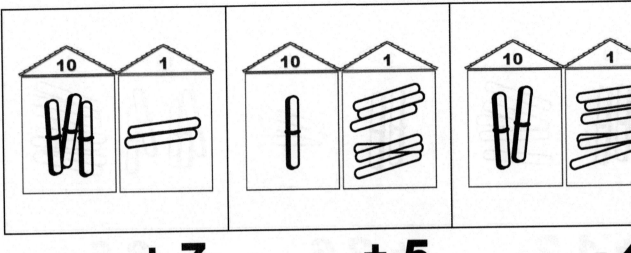

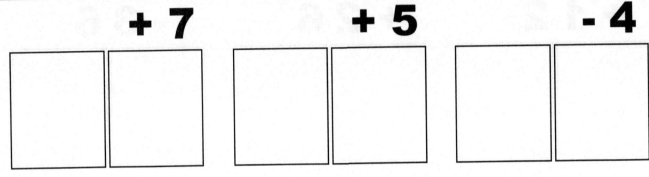

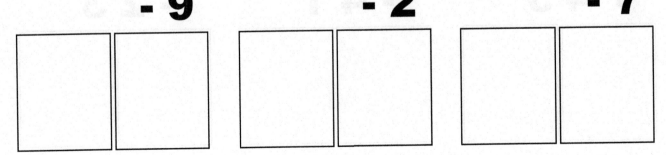

Right-Brained Place Value, 2nd Edition, © 2013 Sarah Major, Child1st Publications, www.child1st.com.

5.1. Computation to 20: a global view

20

20	19	18	17	16	15	14	13	12	11	10
0	1	2	3	4	5	6	7	8	9	10

19

19	18	17	16	15	14	13	12	11	10
0	1	2	3	4	5	6	7	8	9

18

18	17	16	15	14	13	12	11	10	9
0	1	2	3	4	5	6	7	8	9

17

17	16	15	14	13	12	11	10	9
0	1	2	3	4	5	6	7	8

16

16	15	14	13	12	11	10	9	8
0	1	2	3	4	5	6	7	8

15

15	14	13	12	11	10	9	8
0	1	2	3	4	5	6	7

14

14	13	12	11	10	9	8	7
0	1	2	3	4	5	6	7

13

13	12	11	10	9	8	7
0	1	2	3	4	5	6

12

12	11	10	9	8	7	6
0	1	2	3	4	5	6

11

11	10	9	8	7	6
0	1	2	3	4	5

10

10	9	8	7	6	5
0	1	2	3	4	5

Name: _____

11 $+\ 7$	14 $+\ 2$	10 $+\ 4$	12 $+\ 5$	16 $+\ 3$
15 $+\ 1$	13 $+\ 4$	17 $+\ 1$	11 $+\ 3$	15 $+\ 4$
18 $+\ 1$	12 $+\ 7$	10 $+\ 6$	14 $+\ 4$	11 $+\ 4$
16 $+\ 2$	10 $+\ 9$	13 $+\ 5$	17 $+\ 2$	12 $+\ 1$
15 $+\ 3$	11 $+\ 2$	14 $+\ 1$	10 $+\ 1$	13 $+\ 6$

Name: _____

10 + 3	12 + 6	16 + 1	15 + 2	13 + 1
11 + 7	10 + 8	14 + 3	11 + 5	10 + 5
12 + 3	13 + 3	18 + 1	14 + 2	15 + 1
17 + 2	11 + 1	13 + 2	10 + 2	12 + 4
14 + 5	10 + 7	12 + 2	16 + 3	11 + 6

Name: _____

11 + 8	10 + 6	17 + 1	13 + 5	15 + 4
14 + 1	13 + 4	10 + 9	11 + 4	12 + 2
13 + 1	16 + 1	12 + 5	14 + 4	11 + 3
15 + 3	11 + 2	13 + 6	16 + 2	10 + 4
10 + 4	14 + 5	11 + 8	13 + 3	12 + 7

Name: _____

19 - 3	15 - 5	13 - 1	17 - 2	11 - 0
13 - 0	18 - 4	16 - 1	14 - 3	12 - 2
17 - 5	19 - 7	15 - 1	15 - 4	18 - 6
18 - 8	14 - 1	17 - 4	13 - 3	19 - 1
19 - 5	15 - 2	18 - 0	16 - 3	14 - 2

Name: _____

18 - 3	12 - 1	19 - 4	16 - 4	15 - 2
16 - 5	15 - 0	14 - 4	17 - 6	18 - 5
18 - 2	19 - 8	11 - 1	13 - 2	16 - 2
17 - 3	15 - 3	19 - 6	18 - 7	12 - 0
14 - 0	18 - 1	17 - 1	16 - 6	19 - 9

Right-Brained Place Value, 2nd Edition, © 2013 Sarah Major, Child1st Publications, www.child1st.com

Name: _____

18 - 4	15 - 1	11 - 1	19 - 3	13 - 0
16 - 5	17 - 2	14 - 3	15 - 5	16 - 3
19 - 7	12 - 0	18 - 2	13 - 1	17 - 5
15 - 4	13 - 3	19 - 8	18 - 5	14 - 4
16 - 0	18 - 8	17 - 6	12 - 1	19 - 2

6.1. Nines "make a ten" problems

$$\begin{array}{r} 9 \\ + 8 \\ \hline \end{array} \qquad \begin{array}{r} 9 \\ + 3 \\ \hline \end{array} \qquad \begin{array}{r} 9 \\ +9 \\ \hline \end{array} \qquad \begin{array}{r} 2 \\ + 9 \\ \hline \end{array} \qquad \begin{array}{r} 9 \\ + 4 \\ \hline \end{array}$$

$$\begin{array}{r} 9 \\ + 5 \\ \hline \end{array} \qquad \begin{array}{r} 7 \\ + 9 \\ \hline \end{array} \qquad \begin{array}{r} 9 \\ + 6 \\ \hline \end{array} \qquad \begin{array}{r} 9 \\ + 1 \\ \hline \end{array} \qquad \begin{array}{r} 8 \\ + 9 \\ \hline \end{array}$$

- -

6.2. Eights "make a ten" problems

$$\begin{array}{r} 5 \\ + 8 \\ \hline \end{array} \qquad \begin{array}{r} 8 \\ + 6 \\ \hline \end{array} \qquad \begin{array}{r} 8 \\ + 4 \\ \hline \end{array} \qquad \begin{array}{r} 7 \\ + 8 \\ \hline \end{array} \qquad \begin{array}{r} 8 \\ + 3 \\ \hline \end{array}$$

$$\begin{array}{r} 3 \\ + 8 \\ \hline \end{array} \qquad \begin{array}{r} 2 \\ + 8 \\ \hline \end{array} \qquad \begin{array}{r} 8 \\ + 5 \\ \hline \end{array} \qquad \begin{array}{r} 8 \\ + 8 \\ \hline \end{array} \qquad \begin{array}{r} 6 \\ + 8 \\ \hline \end{array}$$

6.3. Eights and nines "make a ten" problems

Name: _____

$$9 + 6 \qquad 5 + 8 \qquad 3 + 8 \qquad 9 + 2 \qquad 5 + 9$$

$$8 + 7 \qquad 8 + 4 \qquad 9 + 9 \qquad 9 + 1 \qquad 6 + 8$$

$$4 + 9 \qquad 9 + 7 \qquad 9 + 3 \qquad 8 + 2 \qquad 8 + 3$$

$$2 + 9 \qquad 7 + 8 \qquad 9 + 8 \qquad 8 + 8 \qquad 6 + 9$$

$$8 + 6 \qquad 9 + 5 \qquad 4 + 8 \qquad 9 + 4 \qquad 8 + 5$$

Right-Brained Place Value, 2nd Edition, © 2013 Sarah Major, Child1st Publications, www.child1st.com.

6.4. Sevens "make a ten" problems

$$\begin{array}{r} 7 \\ + 5 \\ \hline \end{array} \qquad \begin{array}{r} 3 \\ + 7 \\ \hline \end{array} \qquad \begin{array}{r} 7 \\ + 4 \\ \hline \end{array} \qquad \begin{array}{r} 7 \\ + 7 \\ \hline \end{array} \qquad \begin{array}{r} 6 \\ + 7 \\ \hline \end{array}$$

$$\begin{array}{r} 4 \\ + 7 \\ \hline \end{array} \qquad \begin{array}{r} 7 \\ + 7 \\ \hline \end{array} \qquad \begin{array}{r} 7 \\ + 6 \\ \hline \end{array} \qquad \begin{array}{r} 5 \\ + 7 \\ \hline \end{array} \qquad \begin{array}{r} 7 \\ + 3 \\ \hline \end{array}$$

- -

6.5. Sixes "make a ten" problems

$$\begin{array}{r} 6 \\ + 4 \\ \hline \end{array} \qquad \begin{array}{r} 5 \\ + 6 \\ \hline \end{array} \qquad \begin{array}{r} 4 \\ + 6 \\ \hline \end{array} \qquad \begin{array}{r} 6 \\ + 6 \\ \hline \end{array} \qquad \begin{array}{r} 6 \\ + 5 \\ \hline \end{array}$$

$$\begin{array}{r} 6 \\ + 6 \\ \hline \end{array} \qquad \begin{array}{r} 6 \\ + 5 \\ \hline \end{array} \qquad \begin{array}{r} 6 \\ + 4 \\ \hline \end{array} \qquad \begin{array}{r} 5 \\ + 6 \\ \hline \end{array} \qquad \begin{array}{r} 4 \\ + 6 \\ \hline \end{array}$$

Right-Brained Place Value, 2nd Edition, © 2013 Sarah Major, Child1st Publications, www.child1st.com

6.6. Sixes and sevens "make a ten" problems

$$\begin{array}{r} 7 \\ + 6 \\ \hline \end{array} \qquad \begin{array}{r} 6 \\ + 4 \\ \hline \end{array} \qquad \begin{array}{r} 3 \\ + 7 \\ \hline \end{array} \qquad \begin{array}{r} 6 \\ + 5 \\ \hline \end{array} \qquad \begin{array}{r} 7 \\ + 4 \\ \hline \end{array}$$

$$\begin{array}{r} 6 \\ + 6 \\ \hline \end{array} \qquad \begin{array}{r} 7 \\ + 4 \\ \hline \end{array} \qquad \begin{array}{r} 7 \\ + 5 \\ \hline \end{array} \qquad \begin{array}{r} 6 \\ + 4 \\ \hline \end{array} \qquad \begin{array}{r} 5 \\ + 7 \\ \hline \end{array}$$

$$\begin{array}{r} 4 \\ + 6 \\ \hline \end{array} \qquad \begin{array}{r} 6 \\ + 5 \\ \hline \end{array} \qquad \begin{array}{r} 7 \\ + 3 \\ \hline \end{array} \qquad \begin{array}{r} 4 \\ + 7 \\ \hline \end{array} \qquad \begin{array}{r} 7 \\ + 7 \\ \hline \end{array}$$

$$\begin{array}{r} 7 \\ + 4 \\ \hline \end{array} \qquad \begin{array}{r} 7 \\ + 7 \\ \hline \end{array} \qquad \begin{array}{r} 6 \\ + 4 \\ \hline \end{array} \qquad \begin{array}{r} 7 \\ + 6 \\ \hline \end{array} \qquad \begin{array}{r} 7 \\ + 5 \\ \hline \end{array}$$

$$\begin{array}{r} 6 \\ + 7 \\ \hline \end{array} \qquad \begin{array}{r} 7 \\ + 5 \\ \hline \end{array} \qquad \begin{array}{r} 6 \\ + 6 \\ \hline \end{array} \qquad \begin{array}{r} 5 \\ + 6 \\ \hline \end{array} \qquad \begin{array}{r} 7 \\ + 3 \\ \hline \end{array}$$

Name: _____

9 + 4	7 + 6	5 + 8	7 + 7	9 + 2
9 + 8	6 + 6	6 + 4	9 + 7	3 + 8
5 + 7	9 + 3	8 + 7	6 + 9	7 + 4
8 + 8	6 + 5	1 + 9	8 + 2	9 + 5
9 + 6	4 + 8	7 + 3	9 + 9	8 + 6

Name: _____

9 + 8	6 + 4	8 + 5	2 + 9	7 + 6
4 + 7	8 + 8	9 + 7	6 + 5	8 + 3
8 + 2	9 + 1	3 + 7	9 + 9	5 + 7
4 + 9	6 + 6	8 + 4	8 + 7	9 + 5
8 + 6	9 + 6	7 + 5	9 + 3	7 + 7

ght-Brained Place Value, 2nd Edition, © 2013 Sarah Major, Child1st Publications, www.child1st.com.

Name: _____

7 + 3	5 + 9	8 + 6	4 + 6	9 + 2
9 + 9	8 + 3	9 + 4	7 + 7	8 + 4
2 + 8	7 + 4	8 + 7	9 + 1	3 + 9
7 + 5	9 + 6	5 + 6	8 + 3	8 + 8
9 + 8	8 + 5	6 + 6	9 + 7	7 + 6

Right-Brained Place Value, 2nd Edition, © 2013 Sarah Major, Child1st Publications, www.child1st.com

Name: _____

18 $- 9$	14 $- 9$	11 $- 9$	17 $- 9$
13 $- 9$	16 $- 9$	10 $- 9$	12 $- 9$
11 $- 9$	15 $- 9$	14 $- 9$	16 $- 9$
17 $- 9$	12 $- 9$	18 $- 9$	13 $- 9$
14 $- 9$	10 $- 9$	16 $- 9$	15 $- 9$

Name: _____

12	17	15	11
- 8	- 8	- 8	- 8
16	10	14	13
- 8	- 8	- 8	- 8
17	15	10	12
- 8	- 8	- 8	- 8
13	11	17	16
- 8	- 8	- 8	- 8
15	14	13	10
- 8	- 8	- 8	- 8

Name: _____

11 - 8	11 - 9	14 - 8	14 - 9
18 - 8	18 - 9	16 - 8	16 - 9
15 - 8	15 - 9	17 - 8	17 - 9
13 - 8	13 - 9	10 - 8	10 - 9
12 - 8	12 - 9	15 - 8	15 - 9

Name: _____

16 - 9	10 - 8	13 - 8	14 - 9
15 - 8	12 - 9	15 - 8	17 - 9
18 - 9	11 - 8	16 - 8	11 - 9
14 - 8	12 - 9	15 - 9	12 - 8
13 - 9	11 - 8	17 - 8	10 - 9

Right-Brained Place Value, 2nd Edition, © 2013 Sarah Major, Child1st Publications, www.child1st.com

Name: _____

13 - 7	12 - 7	16 - 7	11 - 7
15 - 7	11 - 7	10 - 7	14 - 7
12 - 7	14 - 7	15 - 7	16 - 7
16 - 7	10 - 7	13 - 7	12 - 7
14 - 7	13 - 7	11 - 7	15 - 7

Name: _____

11 − 7	11 − 8	16 − 7	16 − 8
13 − 7	13 − 8	12 − 7	12 − 8
14 − 7	14 − 8	10 − 7	10 − 8
15 − 7	15 − 8	11 − 7	11 − 8
12 − 7	12 − 8	13 − 7	13 − 8

Name: _____

17 - 8	13 - 7	15 - 8	12 - 7
15 - 7	10 - 8	11 - 7	14 - 7
13 - 8	14 - 8	15 - 7	16 - 8
11 - 7	16 - 7	12 - 8	15 - 8
14 - 8	10 - 7	13 - 7	11 - 8

Name: _____

18 - 9	14 - 8	14 - 7	16 - 8	10 - 9
11 - 9	10 - 8	13 - 8	12 - 9	15 - 7
17 - 8	16 - 9	12 - 7	11 - 8	13 - 7
16 - 7	15 - 8	15 - 9	10 - 7	17 - 9
14 - 9	11 - 7	12 - 8	14 - 7	13 - 9

Right-Brained Place Value, 2nd Edition, © 2013 Sarah Major, Child1st Publications, www.child1st.com

Name: _____

15 - 6	11 - 6	14 - 6	12 - 6
10 - 6	13 - 6	12 - 6	15 - 6
14 - 6	11 - 6	10 - 6	13 - 6
10 - 6	12 - 6	14 - 6	11 - 6
13 - 6	15 - 6	10 - 6	12 - 6

Name: _____

14 - 5	12 - 3	10 - 4	11 - 3
11 - 5	11 - 2	13 - 5	13 - 4
12 - 4	10 - 5	12 - 3	10 - 3
11 - 3	10 - 2	11 - 4	12 - 5
11 - 2	12 - 4	14 - 5	10 - 2

Name: _____

18 - 9	12 - 7	13 - 5	14 - 8	15 - 6
11 - 4	11 - 3	14 - 7	13 - 9	10 - 8
12 - 6	14 - 5	12 - 4	10 - 3	11 - 2
10 - 5	16 - 9	15 - 7	11 - 9	14 - 6
13 - 8	11 - 5	11 - 8	13 - 4	17 - 8

Name: _____

17 - 9	13 - 6	12 - 3	15 - 8	15 - 9
12 - 9	11 - 7	16 - 7	10 - 2	14 - 9
10 - 6	12 - 8	18 - 9	13 - 7	11 - 6
14 - 5	16 - 8	10 - 7	10 - 4	15 - 7
11 - 2	13 - 4	13 - 8	12 - 5	10 - 9

Right-Brained Place Value, 2nd Edition, © 2013 Sarah Major, Child1st Publications, www.child1st.com

Name: _____

17 - 8	14 - 8	13 - 5	15 - 6	12 - 4
11 - 5	10 - 3	18 - 9	16 - 9	11 - 3
15 - 8	16 - 7	11 - 8	14 - 7	10 - 5
14 - 6	13 - 9	14 - 5	17 - 9	11 - 9
10 - 8	11 - 4	12 - 6	12 - 7	16 - 8

Name: _____

15 - 9	11 - 6	15 - 7	12 - 5	10 - 9
12 - 3	13 - 8	16 - 9	14 - 7	13 - 4
13 - 7	10 - 6	13 - 5	14 - 9	11 - 7
11 - 2	17 - 8	12 - 8	12 - 4	10 - 4
12 - 9	10 - 7	10 - 2	13 - 6	11 - 4

Right-Brained Place Value, 2nd Edition, © 2013 Sarah Major, Child1st Publications, www.child1st.com.

Name: _____

10 + 5	13 + 2	19 - 5	16 - 1	16 + 1
14 - 2	13 - 3	11 + 3	18 - 5	12 + 7
14 + 4	12 - 1	15 + 2	17 + 1	19 - 6
19 - 9	11 + 6	18 - 7	17 - 5	12 + 4
15 - 3	16 + 3	14 - 0	13 + 5	17 - 4

Name: _____

19 - 3	10 + 1	14 + 3	16 - 6	12 + 2
16 + 2	15 - 4	13 + 4	15 - 0	18 - 3
14 - 3	17 - 0	15 - 2	17 + 2	10 + 7
12 + 5	18 + 1	18 - 1	11 + 8	16 - 3
17 - 7	18 - 2	15 + 3	19 - 7	11 + 1

Name: _____

16 $-\ 5$	10 $+\ 9$	19 $-\ 8$	14 $-\ 4$	15 $+\ 1$
14 $+\ 5$	12 $-\ 0$	13 $+\ 6$	18 $-\ 8$	11 $+\ 2$
12 $+\ 6$	14 $+\ 2$	13 $+\ 3$	13 $-\ 2$	17 $-\ 6$
18 $-\ 4$	10 $+\ 4$	15 $-\ 1$	11 $+\ 5$	19 $-\ 2$
12 $+\ 3$	17 $-\ 3$	16 $-\ 0$	13 $-\ 1$	13 $+\ 1$

Name: _____

17 − 2	10 + 8	16 − 4	19 − 1	12 + 1
14 + 1	18 − 0	11 + 4	14 − 1	10 + 3
12 − 2	15 + 4	19 − 4	16 + 2	13 − 0
10 + 2	16 − 2	12 + 5	17 − 1	11 + 7
18 − 6	18 + 1	10 + 6	15 − 5	19 − 0

Name: _____

19 - 9	10 + 5	16 - 1	12 + 7	18 - 5
16 + 1	17 - 4	14 + 3	15 - 0	11 + 8
11 + 6	13 + 5	17 + 1	19 - 3	14 - 2
13 - 3	12 + 2	16 - 3	15 - 3	10 + 1
15 + 2	18 - 1	13 + 4	17 - 7	12 - 1

Name: _____

13 + 2	19 - 7	14 - 3	12 + 6	10 + 9
17 - 0	14 + 4	16 - 6	18 - 3	16 + 3
11 + 1	17 + 2	15 - 4	15 + 4	12 - 0
15 - 2	13 - 1	10 + 7	19 - 5	11 + 3
12 + 4	18 - 7	16 - 0	13 + 1	17 - 5

Right-Brained Place Value, 2nd Edition, © 2013 Sarah Major, Child1st Publications, www.child1st.com

Name: _____

18 − 9	15 − 9	9 + 1	7 + 9	16 − 9
9 + 2	9 + 3	13 − 9	17 − 9	6 + 9
12 − 9	9 + 5	10 − 9	9 + 9	9 + 4
9 + 8	16 − 9	2 + 9	15 − 9	18 − 9
5 + 9	14 − 9	11 − 9	8 + 9	9 + 6

Name: _____

14 - 9	4 + 9	17 - 9	12 - 9	9 + 7
9 + 5	11 - 9	9 + 9	3 + 9	16 - 9
10 - 9	9 + 2	13 - 9	9 + 6	14 - 9
7 + 9	17 - 9	12 - 9	9 + 1	9 + 4
9 + 8	1 + 9	9 + 3	15 - 9	18 - 9

Right-Brained Place Value, 2nd Edition, © 2013 Sarah Major, Child1st Publications, www.child1st.com.

Name: _____

8 + 6	10 - 8	4 + 8	14 - 8	8 + 7
16 - 8	17 - 8	13 - 8	8 + 5	3 + 8
8 + 2	14 - 8	8 + 8	15 - 8	12 - 8
7 + 8	8 + 3	11 - 8	17 - 8	6 + 8
13 - 8	5 + 8	10 - 8	8 + 4	16 - 8

Name: _____

15 − 8	8 + 2	17 − 8	8 + 5	11 − 8
8 + 7	8 + 8	4 + 8	13 − 8	14 − 8
6 + 8	12 − 8	10 − 8	8 + 3	7 + 8
14 − 8	11 − 8	2 + 8	15 − 8	8 + 4
5 + 8	8 + 6	16 − 8	8 + 8	12 − 8

Name: _____

14 − 7	7 + 6	16 − 7	11 − 7	7 + 3
5 + 7	15 − 7	7 + 4	13 − 7	6 + 7
3 + 7	4 + 7	12 − 7	7 + 7	15 − 7
16 − 7	10 − 7	7 + 5	7 + 6	12 − 7
7 + 7	13 − 7	11 − 7	14 − 7	7 + 4

Right-Brained Place Value, 2nd Edition, © 2013 Sarah Major, Child1st Publications, www.child1st.com.

Name: _____

7 + 5	15 - 7	7 + 7	10 - 7	4 + 7
11 - 7	16 - 7	6 + 7	7 + 3	12 - 7
3 + 7	5 + 7	14 - 7	15 - 7	7 + 6
10 - 7	7 + 4	13 - 7	7 + 5	11 - 7
7 + 7	12 - 7	7 + 3	14 - 7	16 - 7

Right-Brained Place Value, 2nd Edition, © 2013 Sarah Major, Child1st Publications, www.child1st.com.

Name: _____

15 - 6	6 + 6	12 - 6	5 + 6
6 + 5	14 - 6	11 - 6	13 - 6
12 - 6	6 + 4	10 - 6	14 - 6
4 + 6	11 - 6	6 + 5	15 - 6
13 - 6	10 - 6	6 + 6	6 + 4

Name: _____

$$\begin{array}{r} 6 \\ + 5 \\ \hline \end{array} \qquad \begin{array}{r} 12 \\ - 6 \\ \hline \end{array} \qquad \begin{array}{r} 4 \\ + 6 \\ \hline \end{array} \qquad \begin{array}{r} 14 \\ - 6 \\ \hline \end{array}$$

$$\begin{array}{r} 13 \\ - 6 \\ \hline \end{array} \qquad \begin{array}{r} 6 \\ + 6 \\ \hline \end{array} \qquad \begin{array}{r} 10 \\ - 6 \\ \hline \end{array} \qquad \begin{array}{r} 15 \\ - 6 \\ \hline \end{array}$$

$$\begin{array}{r} 14 \\ - 6 \\ \hline \end{array} \qquad \begin{array}{r} 11 \\ - 6 \\ \hline \end{array} \qquad \begin{array}{r} 5 \\ + 6 \\ \hline \end{array} \qquad \begin{array}{r} 13 \\ - 6 \\ \hline \end{array}$$

$$\begin{array}{r} 10 \\ - 6 \\ \hline \end{array} \qquad \begin{array}{r} 6 \\ + 4 \\ \hline \end{array} \qquad \begin{array}{r} 12 \\ - 6 \\ \hline \end{array} \qquad \begin{array}{r} 4 \\ + 6 \\ \hline \end{array}$$

$$\begin{array}{r} 6 \\ + 6 \\ \hline \end{array} \qquad \begin{array}{r} 15 \\ - 6 \\ \hline \end{array} \qquad \begin{array}{r} 11 \\ - 6 \\ \hline \end{array} \qquad \begin{array}{r} 6 \\ + 5 \\ \hline \end{array}$$

Name: _____

14 - 8	9 + 5	15 - 6	11 - 7	7 + 6
6 + 8	13 - 9	8 + 3	3 + 7	17 - 9
9 + 9	16 - 7	12 - 8	6 + 5	10 - 6
10 - 9	8 + 7	11 - 6	6 + 6	15 - 9
7 + 4	12 - 9	9 + 8	13 - 7	9 + 2

Name: _____

8 + 8	10 - 8	6 + 4	14 - 9	12 - 7
15 - 8	9 + 3	8 + 2	12 - 6	4 + 9
13 - 6	11 - 8	9 + 6	7 + 5	17 - 8
5 + 8	14 - 6	18 - 9	8 + 4	15 - 7
9 + 7	9 + 1	16 - 9	10 - 7	7 + 7

Right-Brained Place Value, 2nd Edition, © 2013 Sarah Major, Child1st Publications, www.child1st.com.

Name: _____

11 − 6	8 + 3	14 − 9	5 + 8	6 + 6
7 + 3	11 − 9	16 − 7	9 + 8	13 − 8
4 + 6	12 − 8	7 + 5	14 − 6	8 + 6
8 + 8	10 − 7	9 + 1	16 − 8	17 − 9
12 − 7	7 + 4	10 − 9	4 + 9	14 − 7

Name: _____

$$
\begin{array}{r} 12 \\ -\ 3 \\ \hline \end{array}
\qquad
\begin{array}{r} 4 \\ +\ 8 \\ \hline \end{array}
\qquad
\begin{array}{r} 5 \\ +\ 6 \\ \hline \end{array}
\qquad
\begin{array}{r} 14 \\ -\ 5 \\ \hline \end{array}
\qquad
\begin{array}{r} 11 \\ -\ 5 \\ \hline \end{array}
$$

$$
\begin{array}{r} 8 \\ +\ 2 \\ \hline \end{array}
\qquad
\begin{array}{r} 3 \\ +\ 7 \\ \hline \end{array}
\qquad
\begin{array}{r} 10 \\ -\ 2 \\ \hline \end{array}
\qquad
\begin{array}{r} 4 \\ +\ 9 \\ \hline \end{array}
\qquad
\begin{array}{r} 12 \\ -\ 4 \\ \hline \end{array}
$$

$$
\begin{array}{r} 5 \\ +\ 5 \\ \hline \end{array}
\qquad
\begin{array}{r} 11 \\ -\ 3 \\ \hline \end{array}
\qquad
\begin{array}{r} 9 \\ +\ 3 \\ \hline \end{array}
\qquad
\begin{array}{r} 12 \\ -\ 5 \\ \hline \end{array}
\qquad
\begin{array}{r} 3 \\ +\ 8 \\ \hline \end{array}
$$

$$
\begin{array}{r} 10 \\ -\ 4 \\ \hline \end{array}
\qquad
\begin{array}{r} 13 \\ -\ 5 \\ \hline \end{array}
\qquad
\begin{array}{r} 11 \\ -\ 4 \\ \hline \end{array}
\qquad
\begin{array}{r} 7 \\ +\ 4 \\ \hline \end{array}
\qquad
\begin{array}{r} 5 \\ +\ 7 \\ \hline \end{array}
$$

$$
\begin{array}{r} 5 \\ +\ 9 \\ \hline \end{array}
\qquad
\begin{array}{r} 11 \\ -\ 2 \\ \hline \end{array}
\qquad
\begin{array}{r} 8 \\ +\ 5 \\ \hline \end{array}
\qquad
\begin{array}{r} 13 \\ -\ 4 \\ \hline \end{array}
\qquad
\begin{array}{r} 10 \\ -\ 5 \\ \hline \end{array}
$$

Right-Brained Place Value, 2nd Edition, © 2013 Sarah Major, Child1st Publications, www.child1st.com

Name: _____

5 + 8	11 - 4	9 + 4	12 - 3	14 - 5
4 + 7	2 + 8	12 - 4	5 + 7	4 + 6
10 - 3	13 - 4	9 + 3	13 - 5	10 - 2
5 + 5	12 - 5	11 - 3	3 + 8	6 + 5
11 - 2	4 + 8	3 + 7	2 + 9	10 - 5

Name: _____

7 + 6	12 + 4	9 + 5	16 + 3	8 + 4
13 + 5	11 + 6	6 + 5	14 + 4	14 + 1
10 + 7	9 + 8	7 + 7	13 + 6	8 + 5
9 + 4	12 + 7	11 + 4	10 + 3	7 + 3
8 + 8	13 + 3	12 + 5	9 + 2	10 + 5

Right-Brained Place Value, 2nd Edition, © 2013 Sarah Major, Child1st Publications, www.child1st.com

Name: _____

8 + 6	10 + 9	17 + 2	11 + 3	9 + 6
11 + 8	5 + 5	18 + 1	13 + 2	10 + 8
14 + 3	8 + 3	11 + 7	6 + 4	7 + 5
9 + 7	10 + 4	13 + 4	12 + 6	8 + 7
7 + 4	11 + 5	15 + 4	6 + 6	10 + 1

Name: _____

19 **- 3**	**15** **- 7**	**16** **- 9**	**11** **- 2**	**17** **- 0**
17 **- 7**	**11** **- 8**	**18** **- 1**	**14** **- 3**	**13** **- 4**
11 **- 5**	**10** **- 9**	**17** **- 6**	**12** **- 7**	**19** **- 9**
16 **- 5**	**19** **- 4**	**13** **- 9**	**15** **- 3**	**14** **- 8**
14 **- 1**	**10** **- 6**	**12** **- 2**	**18** **- 5**	**16** **- 1**

Name: _____

16 - 6	10 - 1	13 - 3	11 - 0	18 - 7
14 - 5	19 - 8	12 - 1	15 - 4	17 - 2
13 - 8	10 - 3	16 - 8	18 - 3	12 - 9
17 - 4	15 - 9	11 - 4	14 - 7	19 - 6
12 - 5	18 - 9	10 - 8	13 - 6	15 - 2

Name: _____

13 - 7	13 + 3	17 - 2	15 - 4	14 + 9
10 + 8	11 - 1	17 + 1	17 + 8	12 - 2
12 + 6	18 - 8	14 - 8	13 + 5	10 - 3
16 - 5	15 + 8	16 + 4	11 - 9	18 + 2
11 + 2	19 - 9	15 - 6	14 + 7	12 + 9

Name: _____

15 + 5	10 - 6	17 + 3	12 - 5	15 - 1
13 - 9	11 + 7	14 - 3	19 + 6	16 + 6
12 + 4	18 + 7	13 + 1	17 - 7	11 - 4
17 + 9	15 - 9	16 - 8	10 + 5	19 - 1
11 - 8	19 - 4	14 + 2	18 - 3	12 + 8

Name: _____

44	53	45	64	54
- 22	+ 21	- 24	+ 13	- 32

13	47	55	54	13
+ 25	- 26	+ 24	- 32	+ 33

37	16	59	15	88
- 26	+ 42	- 36	+ 83	- 33

12	46	13	66	62
+ 14	- 24	+ 46	- 33	+ 37

87	84	77	23	75
- 35	+ 15	- 44	+ 46	- 53

Right-Brained Place Value, 2nd Edition, © 2013 Sarah Major, Child1st Publications, www.child1st.com

Name: _____

53 - 23	35 + 21	75 - 24	54 + 15	84 - 72
43 + 24	67 - 46	45 + 23	78 - 37	33 + 33
57 - 24	36 + 41	49 - 36	45 + 43	83 - 52
52 + 47	48 - 26	33 + 46	64 - 32	54 + 35
99 - 35	88 - 15	77 - 44	66 + 23	55 - 53

Name: _____

$$
\begin{array}{r} 434 \\ -\ 222 \\ \hline \end{array}
\qquad
\begin{array}{r} 523 \\ +\ 221 \\ \hline \end{array}
\qquad
\begin{array}{r} 465 \\ -\ 234 \\ \hline \end{array}
\qquad
\begin{array}{r} 654 \\ +\ 123 \\ \hline \end{array}
\qquad
\begin{array}{r} 584 \\ -\ 352 \\ \hline \end{array}
$$

$$
\begin{array}{r} 143 \\ +\ 245 \\ \hline \end{array}
\qquad
\begin{array}{r} 467 \\ -\ 246 \\ \hline \end{array}
\qquad
\begin{array}{r} 152 \\ +\ 237 \\ \hline \end{array}
\qquad
\begin{array}{r} 564 \\ -\ 332 \\ \hline \end{array}
\qquad
\begin{array}{r} 173 \\ +\ 313 \\ \hline \end{array}
$$

$$
\begin{array}{r} 387 \\ -\ 256 \\ \hline \end{array}
\qquad
\begin{array}{r} 136 \\ +\ 442 \\ \hline \end{array}
\qquad
\begin{array}{r} 589 \\ -\ 366 \\ \hline \end{array}
\qquad
\begin{array}{r} 146 \\ +\ 853 \\ \hline \end{array}
\qquad
\begin{array}{r} 848 \\ -\ 333 \\ \hline \end{array}
$$

$$
\begin{array}{r} 222 \\ +\ 437 \\ \hline \end{array}
\qquad
\begin{array}{r} 476 \\ -\ 254 \\ \hline \end{array}
\qquad
\begin{array}{r} 163 \\ +\ 436 \\ \hline \end{array}
\qquad
\begin{array}{r} 656 \\ -\ 333 \\ \hline \end{array}
\qquad
\begin{array}{r} 642 \\ +\ 357 \\ \hline \end{array}
$$

$$
\begin{array}{r} 87 \\ -\ 35 \\ \hline \end{array}
\qquad
\begin{array}{r} 84 \\ +\ 15 \\ \hline \end{array}
\qquad
\begin{array}{r} 77 \\ -\ 44 \\ \hline \end{array}
\qquad
\begin{array}{r} 23 \\ +\ 46 \\ \hline \end{array}
\qquad
\begin{array}{r} 75 \\ -\ 53 \\ \hline \end{array}
$$

Name: _____

4534 - 2422	5333 + 2524	4765 - 2544	6554 + 3245
5343 + 2535	4867 - 2756	5325 + 2344	5984 - 3652
36667 - 24436	23246 + 42532	55649 - 33326	23425 + 63373
32432 + 42357	43646 - 22334	25433 + 44246	67876 - 35653
886567 - 323225	7234324 + 2523225	778887 - 455554	24323 + 42346

Name: _____

$$44568675467894$$
$$- \ 23243224334562$$

$$74673432645362433$$
$$+ \ 23225325244323235$$

$$39786987897687979876897$$
$$- \ 24243524332322345342346$$

$$153454332243452$$
$$+ \ 132443325454324$$

$$853544789757$$
$$- \ 302323343225$$

Right-Brained Place Value, 2nd Edition, © 2013 Sarah Major, Child1st Publications, www.child1st.com

Name: _____

$$14 + 17$$ $$13 + 18$$

$$18 + 25$$ $$27 + 26$$

$$17 + 16$$ $$16 + 47$$

$$19 + 47$$ $$16 + 27$$

$$28 + 35$$ $$77 + 15$$

Name: _____

$$14 + 19$$ $$14 + 18$$

$$29 + 29$$ $$18 + 27$$

$$15 + 29$$ $$18 + 37$$

$$26 + 28$$ $$24 + 37$$

$$23 + 48$$ $$35 + 58$$

Name: _____

$$34 + 57$$ $$23 + 48$$

$$68 + 25$$ $$57 + 27$$

$$37 + 36$$ $$46 + 27$$

$$49 + 47$$ $$56 + 24$$

$$68 + 24$$ $$66 + 25$$

Name: _____

$$34 + 49$$ $$34 + 28$$

$$29 + 67$$ $$48 + 26$$

$$75 + 19$$ $$68 + 35$$

$$46 + 25$$ $$46 + 37$$

$$34 + 48$$ $$37 + 48$$

Name: _____

23 + 37	46 + 28
48 + 25	47 + 26
37 + 46	46 + 47
49 + 47	36 + 28
59 + 35	68 + 25

Name: _____

45 + 29	54 + 38
59 + 28	38 + 47
45 + 29	58 + 37
56 + 29	44 + 38
53 + 39	35 + 57

Name: _____

$$\begin{array}{r} 34 \\ -\ 7 \\ \hline \end{array} \qquad \begin{array}{r} 63 \\ -\ 8 \\ \hline \end{array}$$

$$\begin{array}{r} 78 \\ -\ 9 \\ \hline \end{array} \qquad \begin{array}{r} 53 \\ -\ 6 \\ \hline \end{array}$$

$$\begin{array}{r} 77 \\ -\ 8 \\ \hline \end{array} \qquad \begin{array}{r} 86 \\ -\ 9 \\ \hline \end{array}$$

$$\begin{array}{r} 52 \\ -\ 7 \\ \hline \end{array} \qquad \begin{array}{r} 46 \\ -\ 7 \\ \hline \end{array}$$

$$\begin{array}{r} 92 \\ -\ 7 \\ \hline \end{array} \qquad \begin{array}{r} 77 \\ -\ 8 \\ \hline \end{array}$$

Name: _____

$$\begin{array}{r} 94 \\ -\ 9 \\ \hline \end{array} \qquad \begin{array}{r} 74 \\ -\ 8 \\ \hline \end{array}$$

$$\begin{array}{r} 81 \\ -\ 9 \\ \hline \end{array} \qquad \begin{array}{r} 32 \\ -\ 6 \\ \hline \end{array}$$

$$\begin{array}{r} 55 \\ -\ 9 \\ \hline \end{array} \qquad \begin{array}{r} 84 \\ -\ 7 \\ \hline \end{array}$$

$$\begin{array}{r} 66 \\ -\ 8 \\ \hline \end{array} \qquad \begin{array}{r} 54 \\ -\ 7 \\ \hline \end{array}$$

$$\begin{array}{r} 63 \\ -\ 8 \\ \hline \end{array} \qquad \begin{array}{r} 75 \\ -\ 8 \\ \hline \end{array}$$

Right-Brained Place Value, 2nd Edition, © 2013 Sarah Major, Child1st Publications, www.child1st.com

Name: _____

$$\begin{array}{r} 44 \\ -\ 22 \\ \hline \end{array}$$
$$\begin{array}{r} 14 \\ +\ 17 \\ \hline \end{array}$$

$$\begin{array}{r} 13 \\ +\ 25 \\ \hline \end{array}$$
$$\begin{array}{r} 18 \\ +\ 25 \\ \hline \end{array}$$

$$\begin{array}{r} 37 \\ -\ 26 \\ \hline \end{array}$$
$$\begin{array}{r} 17 \\ +\ 16 \\ \hline \end{array}$$

$$\begin{array}{r} 12 \\ +\ 47 \\ \hline \end{array}$$
$$\begin{array}{r} 19 \\ +\ 47 \\ \hline \end{array}$$

$$\begin{array}{r} 87 \\ -\ 35 \\ \hline \end{array}$$
$$\begin{array}{r} 28 \\ +\ 35 \\ \hline \end{array}$$

Name: _____

$$\begin{array}{r} 523 \\ +\ 221 \\ \hline \end{array}$$
$$\begin{array}{r} 63 \\ -\ 8 \\ \hline \end{array}$$

$$\begin{array}{r} 467 \\ -\ 246 \\ \hline \end{array}$$
$$\begin{array}{r} 53 \\ -\ 6 \\ \hline \end{array}$$

$$\begin{array}{r} 136 \\ +\ 442 \\ \hline \end{array}$$
$$\begin{array}{r} 86 \\ -\ 9 \\ \hline \end{array}$$

$$\begin{array}{r} 476 \\ -\ 254 \\ \hline \end{array}$$
$$\begin{array}{r} 46 \\ -\ 7 \\ \hline \end{array}$$

$$\begin{array}{r} 84 \\ +\ 15 \\ \hline \end{array}$$
$$\begin{array}{r} 77 \\ -\ 8 \\ \hline \end{array}$$

Name: _____

Name: _____

34	63	84	74
- 27	- 28	- 39	- 38
78	53	82	32
- 29	- 26	- 29	- 26
77	86	55	84
- 28	- 27	- 29	- 37
52	46	66	54
- 35	- 37	- 28	- 27
92	77	63	75
- 46	- 28	- 28	- 28

Name: _____

$$\begin{array}{r} 94 \\ -\ 9 \\ \hline \end{array} \qquad \begin{array}{r} 23 \\ +\ 48 \\ \hline \end{array}$$

$$\begin{array}{r} 81 \\ -\ 9 \\ \hline \end{array} \qquad \begin{array}{r} 57 \\ +\ 27 \\ \hline \end{array}$$

$$\begin{array}{r} 55 \\ -\ 9 \\ \hline \end{array} \qquad \begin{array}{r} 46 \\ +\ 27 \\ \hline \end{array}$$

$$\begin{array}{r} 66 \\ -\ 8 \\ \hline \end{array} \qquad \begin{array}{r} 56 \\ +\ 24 \\ \hline \end{array}$$

$$\begin{array}{r} 63 \\ -\ 8 \\ \hline \end{array} \qquad \begin{array}{r} 66 \\ +\ 25 \\ \hline \end{array}$$

Name: _____

$$\begin{array}{r} 74 \\ -\ 38 \\ \hline \end{array} \qquad \begin{array}{r} 45 \\ +\ 29 \\ \hline \end{array}$$

$$\begin{array}{r} 32 \\ -\ 26 \\ \hline \end{array} \qquad \begin{array}{r} 59 \\ +\ 28 \\ \hline \end{array}$$

$$\begin{array}{r} 84 \\ -\ 37 \\ \hline \end{array} \qquad \begin{array}{r} 45 \\ +\ 29 \\ \hline \end{array}$$

$$\begin{array}{r} 54 \\ -\ 27 \\ \hline \end{array} \qquad \begin{array}{r} 56 \\ +\ 29 \\ \hline \end{array}$$

$$\begin{array}{r} 75 \\ -\ 28 \\ \hline \end{array} \qquad \begin{array}{r} 53 \\ +\ 49 \\ \hline \end{array}$$

Name: _____

Name: _____

134 + 177	173 + 178	184 + 189	154 + 178
168 + 275	267 + 275	279 + 269	188 + 267
167 + 156	156 + 477	135 + 299	148 + 377
159 + 457	166 + 267	296 + 248	284 + 357
28 + 35	77 + 15	283 + 478	365 + 588

Right-Brained Place Value, 2nd Edition, © 2013 Sarah Major, Child1st Publications, www.child1st.com

Name: _____

364 - 177	623 - 178
748 - 169	533 - 166
737 - 168	846 - 179
542 - 177	466 - 187
932 - 167	747 - 178

Name: _____

844 - 359	744 - 368
831 - 279	322 - 276
535 - 269	844 - 367
626 - 248	534 - 267
643 - 288	745 - 278

APPENDIX B

MONITORING & TRACKING FORMS

Right-Brained Place Value, 2nd Edition, © 2013 Sarah Major, Child1st Publications, www.child1st.com

Individual Assessment: Computation to Ten (Chapter 2)

Name: _____

Based on seveal observations and samples of written work, draw a line through each problem as it is mastered. When an entire street has been mastered, enter the date of mastery. Note any needs for additional practice in the last column.

Street						Date of Mastery	Extra Practice Needed
Third:	0+3	1+2					
Fourth:	0+4	1+3	2+2				
Fifth:	0+5	1+4	2+3				
Sixth:	0+6	1+5	2+4	3+3			
Seventh:	0+7	1+6	2+5	3+4			
Eighth:	0+8	1+7	2+6	3+5	4+4		
Ninth:	0+9	1+8	2+7	3+6	4+5		
Tenth:	0+10	1+9	2+8	3+7	4+6	5+5	

Individual Assessment: Computation to Ten (Chapter 2)

Name: _____

Based on seveal observations and samples of written work, draw a line through each problem as it is mastered. When an entire street has been mastered, enter the date of mastery. Note any needs for additional practice in the last column.

Street						Date of Mastery	Extra Practice Needed
Third:	0+3	1+2					
Fourth:	0+4	1+3	2+2				
Fifth:	0+5	1+4	2+3				
Sixth:	0+6	1+5	2+4	3+3			
Seventh:	0+7	1+6	2+5	3+4			
Eighth:	0+8	1+7	2+6	3+5	4+4		
Ninth:	0+9	1+8	2+7	3+6	4+5		
Tenth:	0+10	1+9	2+8	3+7	4+6	5+5	

Whole Class Record (Chapter 2)

Check off or date each street when student shows mastery

Name	Third	Fourth	Fifth	Sixth	Seventh	Eighth	Ninth	Tenth

Right-Brained Place Value, 2nd Edition, © 2013 Sarah Major, Child1st Publications, www.child1st.com

Whole Class Record (Chapter 2)

Check off or date each skill when student shows mastery

Name	Dot Cards	Five Frames to 20	Five Frames to 70	My Two Hands	Houses

Right-Brained Place Value, 2nd Edition, © 2013 Sarah Major, Child1st Publications, www.child1st.com.

Whole Class Record (Chapters 3-4)

Check off or date each skill as the student shows mastery

Name	Keeps the correct number of sticks in 1s and 10s places	Makes a ten and places correctly on mat	Takes from ten and places remaining 1s correctly	Subtracts from 1s place before 10s place	Is fluent with method and procedures

Right-Brained Place Value, 2nd Edition, © 2013 Sarah Major, Child1st Publications, www.child1st.com

Progress Report (Chapters 3-4)

Name: _____   **Date** _____

Constructing:	child is still working through the procedure
Emerging:	child uses the correct procedure but is hesitant
Fluent:	child can use the procedure confidently

	Constructing	Emerging	Fluent	Comments:
Keeps the correct number of sticks in 1s and 10s places				
Makes a ten and places correctly on mat				
Takes from ten and places remaining 1s correctly				
Subtracts from 1s place before 10s place				
Is fluent with method and procedures				

Progress Report (Chapters 3-4)

Name: _____   **Date** _____

Constructing:	child is still working through the procedure
Emerging:	child uses the correct procedure but is hesitant
Fluent:	child can use the procedure confidently

	Constructing	Emerging	Fluent	Comments:
Keeps the correct number of sticks in 1s and 10s places				
Makes a ten and places correctly on mat				
Takes from ten and places remaining 1s correctly				
Subtracts from 1s place before 10s place				
Is fluent with method and procedures				

Whole Class Record (Chapters 5-7)

Check off or date each skill as the student shows mastery

Name	Adds 1s	Subtracts 1s	Makes a ten (adding)	Takes from ten (subtracting)	Tells a story demonstrating understanding of symbols

Right-Brained Place Value, 2nd Edition, © 2013 Sarah Major, Child1st Publications, www.child1st.co

Progress Report (Chapters 5-7)

Name: _____   **Date** _____

Constructing:	child is still working through the procedure
Emerging:	child uses the correct procedure but is hesitant
Fluent:	child can use the procedure confidently

	Constructing	Emerging	Fluent	Comments:
Adds 1s				
Subtracts 1s				
Makes a ten (adding)				
Takes from ten (subtracting)				
Tells a story demonstrating understanding of symbols				

Progress Report (Chapters 5-7)

Name: _____   **Date** _____

Constructing:	child is still working through the procedure
Emerging:	child uses the correct procedure but is hesitant
Fluent:	child can use the procedure confidently

	Constructing	Emerging	Fluent	Comments:
Adds 1s				
Subtracts 1s				
Makes a ten (adding)				
Takes from ten (subtracting)				
Tells a story demonstrating understanding of symbols				

Whole Class Record (Chapters 8-9)

Check off or date each skill as the student shows mastery

Name	Tells a story matching the three parts of a problem.	Adds and subtracts 1s in multi-digit numbers	Makes a ten with double-digit numbers	Takes from ten with double-digit numbers	Makes a ten and takes from ten with multi-digit numbers

Progress Report (Chapters 8-9)

Name: _____   **Date** _____

Constructing:	child is still working through the procedure
Emerging:	child uses the correct procedure but is hesitant
Fluent:	child can use the procedure confidently

	Constructing	Emerging	Fluent	Comments:
Tells a story matching the three parts of a problem				
Adds and subtracts 1s in multi-digit numbers				
Makes a ten with double-digit numbers				
Takes from ten with double-digit numbers				
Makes a ten and takes from ten with multi-digit numbers				

Progress Report (Chapters 8-9)

Name: _____   **Date** _____

Constructing:	child is still working through the procedure
Emerging:	child uses the correct procedure but is hesitant
Fluent:	child can use the procedure confidently

	Constructing	Emerging	Fluent	Comments:
Tells a story matching the three parts of a problem				
Adds and subtracts 1s in multi-digit numbers				
Makes a ten with double-digit numbers				
Takes from ten with double-digit numbers				
Makes a ten and takes from ten with multi-digit numbers				

APPENDIX C

ANSWER KEYS

76.

2.7a. 0 / 3 or 3 / 0
1 / 2 or 2 / 1

2.7b. 0 / 4 or 4 / 0
1 / 3 or 3 / 1
2 / 2

77.

2.7c. 0 / 5 or 5 / 0
1 / 4 or 4 / 1
2 / 3 or 3 / 2

2.7d. 0 / 6 or 6 / 0
1 / 5 or 5 / 1
2 / 4 or 4 / 2
3 / 3

78.

2.7e. 0 / 7 or 7 / 0
1 / 6 or 6 / 1
2 / 5 or 5 / 2
3 / 4 or 4 / 3

2.7f. 0 / 8 or 8 / 0
1 / 7 or 7 / 1
2 / 6 or 6 / 2
3 / 5 or 5 / 3
4 / 4

79.

2.7g. 0 / 9 or 9 / 0
1 / 8 or 8 / 1
2 / 7 or 7 / 2
3 / 6 or 6 / 3
4 / 5 or 5 / 4

2.7h. 0 / 10 or 10 / 0
1 / 9 or 9 / 1
2 / 8 or 8 / 2
3 / 7 or 7 / 3
4 / 6 or 6 / 4
5 / 5

85.

2.9a. 2, 0, 3, 3, 1

3, 0, 3, 4, 2, 3,
2, 4

4, 1, 4, 2, 4

1, 1, 4, 4, 3, 3,
0, 3

4, 3, 3, 2, 0

86.

2.9b. 0, 6, 5, 3, 6

1, 2, 5, 6, 4, 6,
5, 5

6, 1, 5, 1, 4

2, 0, 5, 6, 4, 6,
3, 5

6, 4, 5, 2, 3

87.

2.9c. 3, 6, 3, 5, 3

6, 1, 5, 3, 1

2, 2, 1, 5, 0

4, 4, 4, 1, 4

3, 2, 6, 2, 4

88.

2.9d. 4, 2, 8, 8, 1, 7,
6, 0

3, 8, 6, 7, 8

4, 7, 7, 8, 7, 7,
6, 8

6, 2, 7, 1, 0

5, 1, 7, 8, 5, 7,
3, 2

89.

2.9e. 6, 4, 10, 9, 3, 9,
8, 0

4, 10, 8, 9, 10

3, 9, 9, 10, 10,
10, 8, 10

6, 2, 9, 3, 1

7, 2, 9, 10, 6, 9,
1, 5

90.

2.9f. 6, 10, 6, 9, 7

10, 5, 9, 5, 3

1, 4, 2, 9, 5

9, 8, 8, 3, 8

7, 4, 10, 7, 8

100.

3.6. Answers vary

101.

3.7. 13, 25, 36

33, 44, 52

23, 15, 57

41, 20, 4

102.

3.8. Answers vary

136.

4.9a. 39, 23, 22

27, 39, 21

Please Note:

Large number is page number; medium numbers identify activities.

In answers 2.7a-h, / separates the upstairs from the downstairs.

In answers separated by commas, each row of answers represents one row of problems.

We have provided all possible answers for problems 2.7a-h.

137.

    4.9b. 37, 49, 21

          68, 4, 22

138.

    4.9c. 39, 21, 22

          36, 43, 9

140.

    5.2a. 18, 16, 14, 17, 19

          16, 17, 18, 14, 19

          19, 19, 16, 18, 15

          18, 19, 18, 19, 13

          18, 13, 15, 11, 19

141.

    5.2b. 13, 18, 17, 17, 14

          18, 18, 17, 16, 15

          15, 16, 19, 16, 16

          19, 12, 15, 12, 16

          19, 17, 14, 19, 17

142.

    5.2c. 19, 16, 18, 18, 19

          15, 17, 19, 15, 14

          14, 17, 17, 18, 14

          18, 13, 19, 18, 14

          14, 19, 19, 16, 19

143.

    5.3a. 16, 10, 12, 15, 11

          13, 14, 15, 11, 10

          12, 12, 14, 11, 12

          10, 13, 13, 10, 18

          14, 13, 18, 13, 12

144.

    5.3b. 15, 11, 15, 12, 13

          11, 15, 10, 11, 13

          16, 11, 10, 11, 14

          14, 12, 13, 11, 12

          14, 17, 16, 10, 10

145.

    5.3c. 14, 14, 10, 16, 13

          11, 15, 11, 10, 13

          12, 12, 16, 12, 12

          11, 10, 11, 13, 10

          16, 10, 11, 11, 17

146.

    6.1. 17, 12, 18, 11, 13

          14, 16, 15, 10, 17

    6.2. 13, 14, 12, 15, 11

          11, 10, 13, 16, 14

147.

    6.3. 15, 13, 11, 11, 14

          15, 12, 18, 10, 14

          13, 16, 12, 10, 11

          11, 15, 17, 16, 15

          14, 14, 12, 13, 13

148.

    6.4. 12, 10, 11, 14, 13

          11, 14, 13, 12, 10

6.5. 10, 11, 10, 12, 11

      12, 11, 10, 11, 10

149.

    6.6. 13, 10, 10, 11, 11

          12, 11, 12, 10, 12

          10, 11, 10, 11, 14

          11, 14, 10, 13, 12

          13, 12, 12, 11, 10

150.

    6.7a. 13, 13, 13, 14, 11

          17, 12, 10, 16, 11

          12, 12, 15, 15, 11

          16, 11, 10, 10, 14

          15, 12, 10, 18, 14

151.

    6.7b. 17, 10, 13, 11, 13

          11, 16, 16, 11, 11

          10, 10, 10, 18, 12

          13, 12, 12, 15, 14

          14, 15, 12, 12, 14

152.

    6.7c. 10, 14, 14, 10, 11

          18, 11, 13, 14, 12

          10, 11, 15, 10, 12

          12, 15, 11, 11, 16

          17, 13, 12, 16, 13

153.

    7.1. 9, 5, 2, 8

          4, 7, 1, 3

          2, 6, 5, 7

          8, 3, 9, 4

          5, 1, 7, 6

154.

    7.2. 4, 9, 7, 3

          8, 2, 6, 5

          9, 7, 2, 4

          5, 3, 9, 8

          7, 6, 5, 2

155.

    7.3a. 3, 2, 6, 5

          10, 9, 8, 7

          7, 6, 9, 8

          5, 4, 2, 1

          4, 3, 7, 6

156.

    7.3b. 7, 2, 5, 5

          7, 3, 7, 8

          9, 3, 8, 2

          6, 3, 6, 4

          4, 3, 9, 1

157.

    7.4. 6, 5, 9, 4

          8, 4, 3, 7

          5, 7, 8, 9

          9, 3, 6, 5

          7, 6, 4, 8

158.

    7.5a. 4, 3, 9, 8

          6, 5, 5, 4

          7, 6, 3, 2

          8, 7, 4, 3

          5, 4, 6, 5

159.

    7.5b. 9, 6, 7, 5

          8, 2, 4, 7

          5, 6, 8, 8

          4, 9, 4, 7

          6, 3, 6, 3

160.

    7.6. 9, 6, 7, 8, 1

          2, 2, 5, 3, 8

          9, 7, 5, 3, 6

          9, 7, 6, 3, 8

          5, 4, 4, 7, 4

161.

    7.7. 9, 5, 8, 6

          4, 7, 6, 9

          8, 5, 4, 7

          4, 6, 8, 5

          7, 9, 4, 6

162.

    7.8. 9, 9, 6, 8

          6, 9, 8, 9

          8, 5, 9, 7

          8, 8, 7, 7

          9, 8, 9, 8

163.

    7.9a. 9, 5, 8, 6, 9

          7, 8, 7, 4, 2

          6, 9, 8, 7, 9

          5, 7, 8, 2, 8

          5, 6, 3, 9, 9

164.

    7.9b. 8, 7, 9, 7, 6

          3, 4, 9, 8, 5

          4, 4, 9, 6, 5

          9, 8, 3, 6, 8

          9, 9, 5, 7, 1

165.

    7.9c. 9, 6, 8, 9, 8

          6, 7, 9, 7, 8

          7, 9, 3, 7, 5

          8, 4, 9, 8, 2

          2, 7, 6, 5, 8

166.

    7.9d. 6, 5, 8, 7, 1

          9, 5, 7, 7, 9

          6, 4, 8, 5, 4

          9, 9, 4, 8, 6

          3, 3, 8, 7, 7

167.

8.1a. 15, 15, 14, 15, 17

12, 10, 14, 13, 19

18, 11, 17, 18, 13

10, 17, 11, 12, 16

12, 19, 14, 18, 13

168.

8.1b. 16, 11, 17, 10, 14

18, 11, 17, 15, 15

11, 17, 13, 19, 17

17, 19, 17, 19, 13

10, 16, 18, 12, 12

169.

8.1c. 11, 19, 11, 10, 16

19, 12, 19, 10, 13

18, 16, 16, 11, 11

14, 14, 14, 16, 17

15, 14, 16, 12, 14

170.

8.1d. 15, 18, 12, 18, 13

15, 18, 15, 13, 13

10, 19, 15, 18, 13

12, 14, 17, 16, 18

12, 19, 16, 10, 19

171.

8.1e. 10, 15, 15, 19, 13

17, 13, 17, 15, 19

17, 18, 18, 16, 12

10, 14, 13, 12, 11

17, 17, 17, 10, 11

172.

8.1f. 15, 12, 11, 18, 19

17, 18, 10, 15, 19

12, 19, 11, 19, 12

13, 12, 17, 14, 14

16, 11, 16, 14, 12

173.

8.2a. 9, 6, 10, 16, 7

11, 12, 4, 8, 15

3, 14, 1, 18, 13

17, 7, 11, 6, 9

14, 5, 2, 17, 15

174.

8.2b. 5, 13, 8, 3, 16

14, 2, 18, 12, 7

1, 11, 4, 15, 5

16, 8, 3, 10, 13

17, 10, 12, 6, 9

175.

8.3a. 14, 2, 12, 6, 15

8, 9, 5, 13, 11

10, 6, 16, 7, 4

15, 11, 3, 9, 14

5, 13, 2, 12, 8

176.

8.3b. 7, 10, 9, 13, 3

15, 16, 12, 5, 6

14, 4, 2, 11, 15

6, 3, 10, 7, 12

13, 14, 8, 16, 4

177.

8.4a. 7, 16, 9, 4, 10

12, 8, 11, 6, 13

10, 11, 5, 14, 8

9, 3, 12, 13, 5

14, 6, 4, 7, 11

178.

8.4b. 12, 8, 14, 3, 11

4, 9, 13, 10, 5

10, 12, 7, 8, 13

3, 11, 6, 12, 4

14, 5, 10, 7, 9

179.

8.5a. 9, 12, 6, 11

11, 8, 5, 7

6, 10, 4, 8

10, 5, 11, 9

7, 4, 12, 10

180.

8.5b. 11, 6, 10, 8

7, 12, 4, 9

8, 5, 11, 7

4, 10, 6, 10

12, 9, 5, 11

181.

    8.6a. 6, 14, 9, 4, 13

        14, 4, 11, 10, 8

        18, 9, 4, 11, 4

        1, 15, 5, 12, 6

        11, 3, 17, 6, 11

182.

    8.6b. 16, 2, 10, 5, 5

        7, 12, 10, 6, 13

        7, 3, 15, 12, 9

        13, 8, 9, 12, 8

        16, 10, 7, 3, 14

183.

    8.6c. 5, 11, 5, 13, 12

        10, 2, 9, 17, 5

        10, 4, 12, 8, 14

        16, 3, 10, 8, 8

        5, 11, 1, 13, 7

184.

    8.7a. 9, 12, 11, 9, 6

        10, 10, 8, 13, 8

        10, 8, 12, 7, 11

        6, 8, 7, 11, 12

        14, 9, 13, 9, 5

185.

    8.7b. 13, 7, 13, 9, 9

        11, 10, 8, 12, 10

        7, 9, 12, 8, 8

10, 7, 8, 11, 11

9, 12, 10, 11, 5

186.

    8.8a. 13, 16, 14, 19, 12

        18, 17, 11, 18, 15

        17, 17, 14, 19, 13

        13, 19, 15, 13, 10

        16, 16, 17, 11, 15

187.

    8.8b. 14, 19, 19, 14, 15

        19, 10, 19, 15, 18

        17, 11, 18, 10, 12

        16, 14, 17, 18, 15

        11, 16, 19, 12, 11

188.

    8.9a. 16, 8, 7, 9, 17

        10, 3, 17, 11, 9

        6, 1, 11, 5, 10

        11, 15, 4, 12, 6

        13, 4, 10, 13, 15

189.

    8.9b. 10, 9, 10, 11, 11

        9, 11, 11, 11, 15

        5, 7, 8, 15, 3

        13, 6, 7, 7, 13

        7, 9, 2, 7, 13

190.

    8.10a. 6, 16, 15, 11, 23

18, 10, 18, 25, 10

18, 10, 6, 18, 7

11, 23, 20, 2, 20

13, 10, 9, 21, 21

191.

    8.10b. 20, 4, 20, 7, 14

        4, 18, 11, 25, 22

        16, 25, 14, 10, 7

        26, 6, 8, 15, 18

        3, 15, 16, 15, 20

192.

    9.1a. 22, 74, 21, 77, 22

        38, 21, 79, 22, 46

        11, 58, 23, 98, 55

        26, 22, 59, 33, 99

        52, 99, 33, 69, 22

193.

    9.1b. 30, 56, 51, 69, 12

        67, 21, 68, 41, 66

        33, 77, 13, 88, 31

        99, 22, 79, 32, 89

        64, 73, 33, 89, 2

194.

    9.1c. 212, 744, 231, 777, 232

        388, 221, 389, 232, 486

        131, 578, 223, 999, 515

(Continued on p. 222)

659, 222, 599, 323, 999

52, 99, 33, 69, 22

195.

9.1d. 2112, 7857, 2221, 9799

7878, 2111, 7669, 2332

12231, 65778, 22323, 86798

74789, 21312, 69679, 32223

563342, 9757549, 323333, 66669

196.

9.1e. 21325451133332

97898757889685668

155434635653656336364 2651

285897657697776

551221446532

197.

9.2a. 31, 31

43, 53

33, 63

66, 43

63, 92

9.2b. 33, 32

58, 45

44, 55

54, 61

71, 93

198.

9.2c. 91, 71

93, 84

73, 73

96, 80

92, 91

9.2d. 83, 62

96, 74

94, 103

71, 83

82, 85

199.

9.2e. 60, 74

73, 73

83, 93

96, 64

94, 93

9.2f. 74, 92

87, 85

74, 95

85, 82

92, 92

200.

9.3a. 27, 55

69, 47

69, 77

45, 39

85, 69

9.3b. 85, 66

72, 26

46, 77

58, 47

55, 67

201.

9.4a. 22, 31

38, 43

11, 33

59, 66

52, 63

9.4b. 744, 55

221, 47

578, 77

222, 39

99, 69

202.

9.5a. 7, 35

49, 27

49, 59

17, 9

46, 49

9.5b. 45, 36

53, 6

26, 47

38, 27

35, 47

203.

9.6a. 85, 71

72, 84

46, 73

58, 80

55, 91

9.6b. 36, 74

6, 87

47, 74

27, 85

47, 102

204.

9.7a. 311, 351

443, 542

323, 633

616, 433

63, 92

9.7b. 373, 332

548, 455

434, 525

544, 641

761, 953

205.

9.8a. 187, 445

579, 367

569, 667

365, 279

765, 569

9.8b. 485, 376

552, 46

266, 477

378, 267

355, 467

CPSIA information can be obtained
at www.ICGtesting.com
Printed in the USA
BVOW10s0732120118
504943BV00005B/18/P